Portable Antennas for Everyone

Compiled and edited by
Steve Telenius-Lowe, PJ4DX

Radio Society of Great Britain

Published by the Radio Society of Great Britain, 3 Abbey Court, Fraser Road, Priory Business Park, Bedford MK44 3WH. Tel: 01234 832700. Web: www.rsgb.org

Published 2020.

ISBN: 9781 9101 9385 3

Cover design: Kevin Williams, M6CYB

Design and layout: Steve Telenius-Lowe, PJ4DX

Production: Mark Allgar, M1MPA

Printed in Great Britain by Hobbs The Printers Ltd. of Totton, Hampshire

Any amendments or updates to this book can be found at www.rsgb.org/booksextra

Contents

Preface

Many years ago I worked for the BBC in Broadcasting House, London. There, the engineering department's definition of "portable" was anything that had handles on it. This definition encompassed massive 15 i.p.s. tape recorders on which we would edit pre-recorded programmes for broadcast and which, although they required two people to lift them, were considered to be portable units – just because they had handles on the side.

I was reminded of this when I tried to come up with a suitable definition of what could be considered a portable antenna. Having taken part in some big DXpeditions – which are surely just portable 'field day' type operations, except from overseas – in which our antennas included multi-element HF Yagis and verticals up to 27m (88ft) high, I realised that the definition of a portable antenna could include almost anything that will radiate.

However, a full-size 3- or 4-element Yagi for 20m or a '4-Square' array of four phased quarter-wave verticals for 80m do not conform to most people's idea of a portable antenna so, for the purposes of this book, I have selected only those antennas that were designed specifically for portable operations. This might be because they are smaller or lighter than permanent 'fixed' or base station antennas or perhaps because they were designed to be assembled and erected quickly and easily.

Activities where there is a need for portable antennas under this definition include field day operations such as NFD (National Field Day) and SSB Field Day in the UK (or ARRL Field Day in North America), club activities such as International Marconi Day or the International Lighthouse Lightship Weekend (ILLW), operating while on holiday from a temporary location whether in the UK or overseas, full-blown DXpeditions, and what might best be described as "OTA operations" such as JOTA (Jamboree on the Air), IOTA (Islands on the Air) and SOTA (Summits on the Air).

With ever-increasing noise levels caused by ADSL, plasma screen television sets, LED ceiling lights or 'wall wart' switch-mode power supplies making operating from home less pleasant if not totally impossible, portable operating has never been as popular as it is today. I am lucky that, living in Bonaire, generally speaking my local noise level is relatively low. But it only requires one neighbour innocently to fit a single dodgy Chinese-made LED ceiling light in his house to completely wipe out the spectrum from about 4.5 to 23MHz, as happened recently. The light radiated an S9+ noise that was audible on a portable receiver nearly a hundred metres away and, had

I been living in a built-up area, it might have been nearly impossible to trace the source.

If you are suffering from high domestic noise levels at home, operating portable from the countryside can be a real revelation. Often there is, in effect, no local noise whatsoever and you find to your surprise that you are copying S1 signals perfectly and can hear the whine of the mosquitoes in the operator's shack in Port Moresby. I exaggerate, but you get the point...

In these pages you will find numerous ingenious designs of antennas intended specifically for portable operating. Our thanks go to the many amateurs who designed them and took the time and effort to write up their projects. Because many amateurs choose to use commercially-made antennas for their portable operations, there is also a chapter of reviews of commercially available antennas that were designed for portable use.

I would like to thank Giles Read, G1MFG, the *RadCom* Technical Editor, and Steve Ford, WB8IMY, ARRL Publications Manager and *QST* Editor, for their help in supplying the material used to compile this book.

Now there is no excuse: with this book in hand this really is the time to get on the air while experiencing the great outdoors!

Steve Telenius-Lowe, PJ4DX
Bonaire, February 2020

Chapter 1

Fibreglass Poles

BUILDING BLOCKS: FIBREGLASS ANTENNA SUPPORT POLES

Often erroneously called 'fishing poles', these poles made specifically for supporting antennas are in fact considerably stronger than those actually designed for catching fish. (I once almost bought a 9m-long 'roach pole' from a fishing supplies store but the shop assistant, in an attempt to demonstrate how strong the pole was, extended it and leant against it, whereupon it promptly snapped in two.) The fibreglass poles marketed by Spiderbeam **[1]** (sold in the UK by Nevada Radio **[2]**) or DX-Wire **[3]** from Germany, or SOTABEAMS **[4]** and DX Commander **[5]** in the UK, are made of far stronger stuff and are ideal as the basic building blocks for numerous portable antenna designs.

Fibreglass antenna support poles come in sizes from 6m (19.6ft) long (e.g. from SOTABEAMS) and costing a modest few tens of pounds up to a whopping 26m (85ft) long (e.g. from Spiderbeam) – which costs several hundred pounds. The advantage of the shorter SOTABEAMS poles is that they collapse down to a smaller length so that they can fit into a rucksack or suitcase – ideal for SOTA (Summits On The Air) activity or 'holiday' operations from abroad – whereas the 10m and 12m poles are generally around 1.20m long when fully collapsed and the bigger ones longer still. Except when you need a really efficient vertical radiator for 160 or 80m, the most useful poles are probably the 10m or 12m ones.

Once extended, the sections are held in place by hose clamps covered in rubber, to avoid damaging the fibreglass. Other than for the quickest of portable operations avoid the temptation to just use plastic insulating tape: it rarely holds for more than a few hours and leaves behind a nasty sticky residue that can be difficult to remove.

Use hose clamps backed by rubber to secure each section of a fibreglass telescopic antenna support pole.

A 40m quarter-wave vertical is an efficient DX antenna, assuming you have a suitable radial system beneath it, and such an antenna can be fitted on a 10m-long fibreglass pole with the radials lying on the ground: perfect for portable operating. I recommend wrapping one turn of wire around each section of the pole and securing the wire to the pole with insulating or (better, because it is less sticky) self-amalgamating tape to reduce the amount that the wire flaps against the pole when it bends in the wind. This also adds a small amount of loading, meaning the vertical height required can be reduced slightly (useful as a quarter-wave wire at 7100kHz is slightly longer than 10.0m). It also approximates to a 'fat' vertical, which increases the bandwidth of the antenna slightly.

A 40m quarter-wave vertical built on a 12m long fibreglass pole. The four radials are elevated about 2m above ground.

An alternative for a 40m quarter-wave vertical is to use a 12m-long pole, in which case the feed point becomes about 2m above ground and a smaller number of elevated resonant quarter-wave long radials can be used, thus saving valuable setting-up time in a portable environment when you just want to get on the air as quickly as possible (though you will then need suitable places to which to tie off the ends of the radials). A minimum of two elevated radials, in a straight line (at 180° to each other) are all that is required, though four (every 90°) would be better, and more better still. Unlike with ground radials, though, where "the more the merrier" is the catch-phrase, it's unlikely that increasing elevated radials beyond 10 or 12 will increase performance further.

If intending to use a fibreglass mast for supporting an inverted-V dipole, first remove the spindly top section by taking it out from the bottom of the pole before it is extended.

While best suited for supporting vertical antennas, these fibreglass poles can also be used to support inverted-V dipoles if a couple of simple precautions are first taken. No matter what length of pole is used, the very top section will almost certainly be too spindly to be used to support a horizontal antenna and it should therefore be removed by taking it out from the bottom of the pole before it is extended (and, in some cases, the second section may also need to be removed).

Use lightweight wire to make the antenna and don't use a balun at the feed point as its additional weight will be too much for the pole. Purists will undoubtedly object to this but the dipole will radiate perfectly well without a balun (and bear in mind we are discussing antennas for temporary portable operation and not a permanent home station set-up). If you do still wish to use a balun you could locate it at the bottom of the pole, or towards the bottom where it is far sturdier. (It is likely that the feeder above the balun would then also

A 12m fibreglass mast used to support a 40m inverted-V dipole. The top section of the pole has been removed to make an approximately 11m high support. It can be seen how much the pole is bending in the wind. Note that lightweight wire is used for the dipole, there is no balun at the feed point, and the coaxial cable has been wrapped around the mast to prevent its weight causing the pole to bend even further.

radiate, though this is not necessarily a bad thing in itself and it may even improve the low-angle radiation. The overall antenna system is also likely to become more omni-directional than a normal dipole or inverted-V.)

Do not allow the coaxial cable to drop down vertically from the feed point as the weight of the coax will cause the pole to bend even further. Instead, wrap one turn of coax around each section of the pole so that it is balanced around the pole and its entire weight is supported vertically (see photo above). The pole will still bend in the wind and with the weight of the antenna wire, but unless it has been really overloaded this is nothing to be concerned about as the fibreglass masts are designed to bend without breaking. (However, to repeat: particularly when supporting horizontal antennas do use proper fibreglass poles designed for supporting antennas, and not the cheaper fishing rods which might well break with too much weight on top.)

The ends of the inverted-V should be tied off to suitable fixed points and this can also help the stability of the pole. Crossed inverted-V dipoles, with a wire every 90º, would be better still as far as stability is concerned and should give almost omni-directional coverage, though I've not tried this.

On the following pages is an independent review of one of the fibreglass antenna support poles.

SPIDERPOLE TELESCOPING ANTENNA MAST

Reviewed by Joel R Hallas, W1ZR
QST, February 2010

Spiderbeam is a German company that is perhaps best known for their lightweight Spiderbeam Yagis. At the heart of the mechanical design of their Yagis are fibreglass poles used to support the wire elements. The fibreglass poles are also available as stand-alone products.

This review is of their 40 foot (12 metre) 'Spiderpole'. If you order a Spiderpole, that's what you get – no more, no less. Our Spiderpole came as a single lightweight package (see **Table 1.1** and the photo left). The pole weighs only 7 pounds (3.3kg) and the 12 sections, fully collapsed, make a handy transport package less than four feet long. The construction is of multilayer fibreglass tubing with thick walls. Actually you also get two rubber end caps that fit over the ends of the largest diameter piece and keep the inner sections from falling out as you handle the collapsed mast.

The 40 foot Spiderpole fibre-glass mast as it comes from the box.

Each section is tapered, so they only function properly if extended in one direction. As each section is extended, a friction fit holds it in place. The idea is that the friction fit, perhaps augmented by a wrap of electrical tape, is sufficient for temporary use, while an optional clamp kit is offered for more permanent installations. In my experiments, the friction fit did hold, but just long enough to allow some additional action. Tape is recommended to secure the joints for portable use and would likely work fine. In an unscheduled experiment, I found that, without tape, if one section came down the shock caused all sections to come down in a rather dramatic but benign way.

We opted for the clamp kit. The clamp kit includes the parts needed to fabricate 11 padded stainless steel clamps, designed to fit above each joint of the extended sections. A bit of fabrication is required, and fortunately the

Extended length (height):	12 m (40ft)
Transportation length:	1.18 m (3ft 10in)
Weight:	3.3kg (7lb)
Bottom diameter:	55mm (2-1/6in)
Top diameter:	8mm (1/3in)
Wall thickness:	1.4mm to 2mm (1/18in to 1/12in)
Number of sections:	12
Pole material:	Black fibreglass, UV protected, reinforced multilayer winding.

Table 1.1: Spiderpole heavy duty telescoping mast, manufacturer's specifications (verified by ARRL).

clamp kit does include instructions. The kit includes lengths of rubber padding and shrink tubing. You first cut the bulk material into sections specified (in mm) in the instructions. The clamps are then straightened, the padding is added, and the tubing is pushed over the straightened clamp and padding to hold everything together. A heat gun (e.g. hair dryer) is used to tighten the tubing.

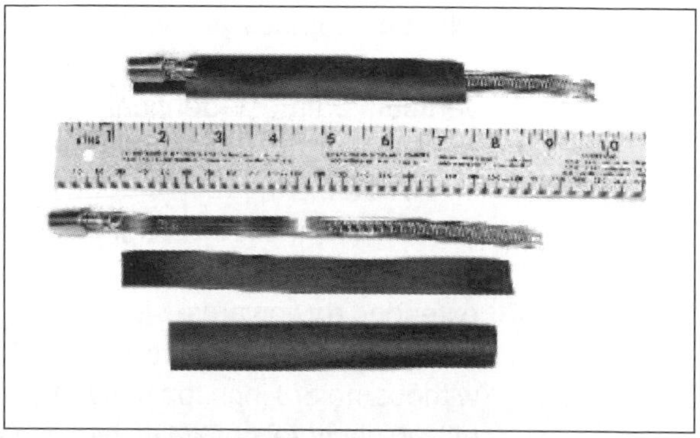

The clamp kit in the process of assembly.

The hardest part was getting the clamps back together. Eventually they all yielded to a pair of pliers and a 7mm socket driver. Fortunately, unlike US-sized hose clamps, all of the sizes of clamp used the same size wrench. A hex head nut driver is my first choice for this kind of operation, but I don't seem to have a metric set.

I was interested to see if raising this mast to full height was really a one person operation. First, I pre-installed a clamp on each section, just snug enough not to fall off. I used the clamp to start each section headed skywards. I found I had better leverage by extending the second section about a foot to move operations closer to shoulder level. To keep the bottom from moving, I found a garden trowel with a handle that made a nice fit into the bottom section and stuck the blade in the soil to its hilt. Up she went, one section after another, with the clamps slid to the bottom of each section and tightened as I went. The last step was to push up the second section to its full height. Working by myself, the mast went up quickly and without difficulty – no surprises, just as advertised (see photo).

The mast is up! The fully extended mast is surprisingly rigid, probably due to the thick wall sections and multilayer fibreglass lay-up.

I was glad I had the clamp kit, however. Trying to do it by myself while wrapping tape around the joints would not have been fun. A hint: if you are by yourself, don't drop the wrench. There just aren't enough hands available to easily lean over and pick up the wrench while keeping the mast upright. Another caution – while pushing up the tapered sections prior to tightening the clamp, be careful not to lift the whole mast off the centre support. I did that once, as an unscheduled experiment, and what had been an easily controllable situation rapidly became a real handful.

This is really a review of a mast, but it seems sensible to briefly discuss antennas that might be used with it.

Obvious choices are vertical wires and inverted-V dipoles. Verticals are perhaps the easiest, except for the need for a ground system, and a 40 foot vertical is almost identical to the popular 43 foot height. A perpendicular pair of inverted-Vs for different bands, tied together in parallel and fed with small coax, can provide double service by serving as guy wires for the top section. A small VHF Yagi could easily be installed lower on the mast, or a small VHF vertical could be mounted on top along with wire HF antennas extended downwards.

If the mast construction is too light for a given antenna, consider using it without the 3/8 inch top section extended. At around 34 feet of height, we're talking diameters of 3/4 inches or more, and a 20m half-wavelength horizontal antenna can be quite effective at that height.

While the mast with a vertical wire may be sufficiently sturdy to support itself, I would be inclined to use at least one set of guys at perhaps the 25 to 30 foot level. They could be easily tied above a clamp. I'd consider two sets a good idea for other antenna types, with inverted-Vs acting as the upper set if used. Spiderbeam offers rolls of 2mm diameter Kevlar guy line, or you could use light rope or heavy duty fishing line.

For a field day or emergency communications event two mounting options come to mind, and your circumstances will likely generate more ideas:

- If there is a sturdy fence post or railing support, the Spiderpole could be temporarily attached using large size clamps [or cable ties – *Ed*], or even rope if there's nothing better at hand.
- Guying is the usual support method of choice for such a mast. Use a centre support, such as my trowel, or better yet a section of steel pipe driven solidly into the soil. My suggestion is to measure and deploy the guys in advance so that they become tight as the mast rises. By tying the guys above the optional clamps and driving their stakes in ahead of raising the mast, they will tighten automatically at just the right point. This can be set up using trigonometry, or by laying each pair of guys on the ground with the antenna temporarily extended along the ground. Large tent stakes are one possibility as guy anchors for temporary use, or local objects of opportunity can be pressed into service.

The bottom line: The Spiderpole mast provides a very usable and strong non-metallic antenna support for portable, emergency communications or even home station use. Once the clamps are set up, the mast can be raised and lowered in a very controlled way with minimal effort.

REFERENCES

[1] Spiderbeam: spiderbeam.com (or, in North America, spiderbeam.us)
[2] Nevada: nevadaradio.co.uk
[3] DX-Wire: dx-wire.de
[4] SOTABEAMS: sotabeams.co.uk
[5] DX Commander: m0mcx.co.uk

Chapter 2

HF Verticals

THE 'WONDER WHIP'

John Goody, M1IOS
RadCom, June 2006

The 'Whizz Whip' is a commercially-made antenna which is described in the relevant chapter in this book. But if you are more into 'homebrewing' rather than using a commercially-made antenna, it is perfectly possible to make something very similar yourself, as demonstrated by John Goody, M1IOS. He called it the 'Wonder Whip' but acknowledged that he based his design on an earlier article by Robert Victor, VA2ERY **[1]**.

John's Wonder Whip uses an autotransformer to transform impedances in order to match a short telescopic whip over a wide range of frequencies for portable QRP operation. The autotransformer is a double-wound transformer; the bottom, primary part being connected to the rig, whilst the secondary is connected to the whip. The impedance transformation is the square of the ratio between the secondary and primary windings. The tapping point varies the ratio between the secondary and primary windings and establishes an impedance transformation.

The original VA2ERY design used a continuously-variable autotransformer based on a 60-turn coil with the impedance ratio selected by a wiper. However, to simplify the design, M1IOS used incremental, rather than continuously variable tuning.

The circuit is shown in **Fig 2.1**. The design uses a 36mm OD ferrite ring wound with 50 turns of 16-gauge antenna wire. The core is tapped every second turn. Two 12-way rotary switches were used, the 12th contact of the first rotary switch is connected to the pole of the second switch, to give

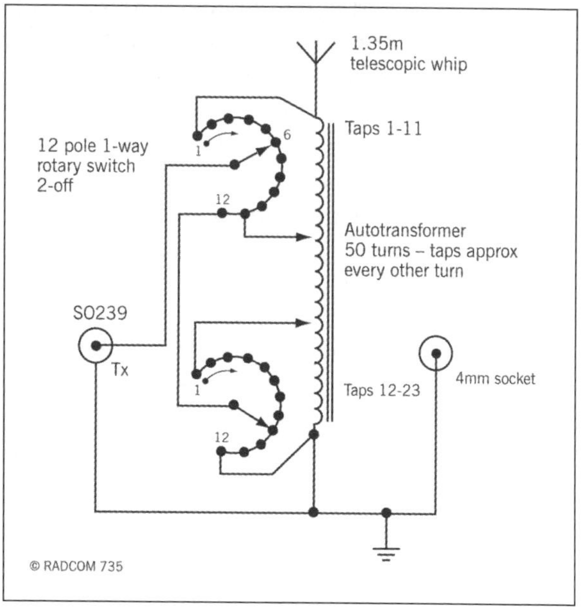

Fig 2.1: The circuit diagram of the M1IOS 'Wonder Whip'.

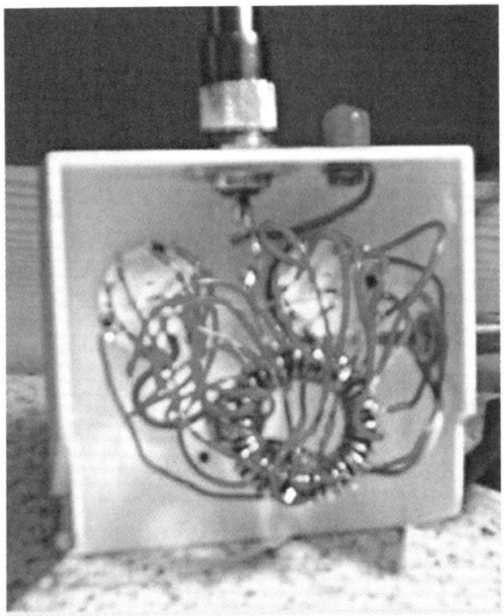

The inside and outside of the Wonder Whip built by John Goody, M1IOS.

23 positions of adjustment. The 23 wire connectors to the toroid are directly soldered to the windings. This was achieved by cleaning and tinning the relevant core winding, forming a small tinned loop to each connecting wire, positioning each wire on the tinned core winding and soldering the two tinned surfaces to fuse into a secure joint.

The telescopic whip was a 10-section antenna 1.31m (4ft 3.5in) long, and 10mm diameter at the base. It has a 4mm hole to accommodate the antenna wire. A connector was made by drilling out the cable entry of a PL-259 connector to a 10mm interference fit into which the telescopic whip was screwed. The whip is secured to the PL-259 connector by soldering the antenna wire into position.

The antenna is tuned by selecting the required band on the radio, and then rotating the two rotary switches on the ATU until the highest background noise or signal is achieved. The antenna peaks on receive and provides a good starting point for transmitting. Set your rig to a low power setting (e.g. 5W) and transmit while observing the SWR meter. Rotate the rotary switch knobs until the SWR is minimised. Fine adjustments to the SWR may be made by altering the length of the telescopic whip (but *not* whilst transmitting!)

A 10 – 12m (33 – 40ft) length of 1mm-diameter wire is used as a counterpoise, which simply lies on the ground.

For VHF and UHF operation the first position on the control knob is selected and the antenna adjusted to a quarter wavelength. For 2m SSB operation horizontal polarisation is achieved by simply turning the antenna on its side!

On receive, performance is first-rate and the S-meter readings appear to be very respectable.

Band	No counterpoise	With counterpoise
10m	1:1	1:1
12m	1:1	1:1
15m	2:1	1:1
17m	1:1	1:1
20m	1.5:1	1:1
30m	1.5:1	1:1
40m	4:1	2:1
80m	8:1	6:1

Table 2.1: SWR figures measured on transmit, with and without the counterpoise wire. On 10m and 12m the telescopic whip length was reduced for minimum SWR.

The best SWR figures that could be achieved using the 1.31m whip, both without and with a 12m counterpoise wire, are shown in **Table 2.1**.

On transmit, some reports given on 40m to M1IOS/P included a 53 from GB2IWM, SM4YPG gave a 47 to 57 report with just 5W; while on 21MHz AC5N in Oklahoma gave M1IOS/P a 59 report when he was running 15W to the Miracle Whip.

THE G0KYA END-FED HALF-WAVE ANTENNA

Steve Nichols, G0KYA
RadCom, October 2010

The use of end-fed half-wave (EFHW) vertical wire antennas is popular in continental Europe, particularly in the Netherlands where there is a commercial manufacturer of such antennas, but it seems they have never really taken off in the UK. Steve Nichols, G0KYA, however, designed and built an end-fed half-wave vertical wire antenna for portable operations by the Norfolk Amateur Radio Club.

Steve explained that he liked to use half-wave dipoles but, for portable operations, the 7m and 10m fibreglass poles he had – while great for supporting verticals – did not have sufficient lateral strength to support horizontal dipoles. Steve said, “What I really wanted to do was have a vertical half-wave dipole, but the problem is that while the impedance at the centre of a dipole is about 50 – 75Ω, and very easy to match to coax, an end-fed half-wave has a very high impedance indeed, around 2000 – 3000Ω. If you just connect it to your coax or rig you will be disappointed. What can you do? After a lot of searching on the web I found the answer. Steve, AA5TB, of Fort Worth, Texas has a great site **[2]** with lots of information.”

G0KYA originally built an end-fed half-wave for 10m. First take a half-wave length of wire. Using the formula 468/frequency this works out at 16ft

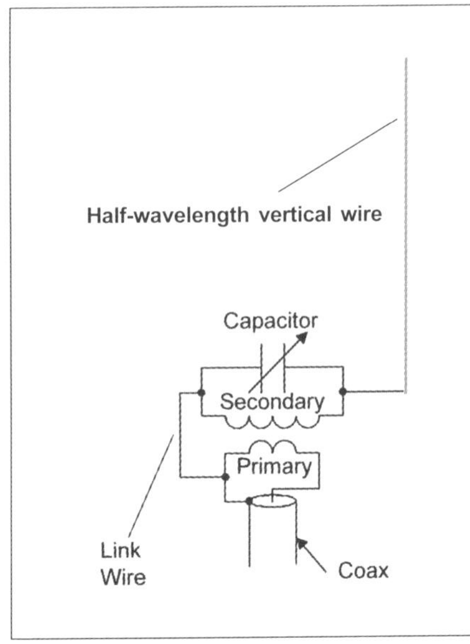

Fig 2.2: The G0KYA end-fed half-wave vertical wire antenna, showing the matching network. The capacitor is a short length of RG-58 coax (see text).

5in (precisely 5m). Next you need a T200 (red) toroid. These cost £4 from JAB Electrical Components **[3]** and can often be found at rallies. G0KYA wound 17 turns of enamelled copper wire on the toroid as the secondary winding – each time the wire passes through the toroid is considered one turn. The wire he used was about 1.25mm in diameter (18 SWG) and this can also be supplied by JAB. Leave a little at the end for connections and then wind two turns over this for the primary, again leaving a little spare.

Across the 17-turn winding you need to connect a small capacitor. RG-58 coax has a capacitance of about 28.8pF per foot: cut off about 10in (25 – 26cm) and connect that across the ends of the winding. An electrical connector ('choc block') was used for the connections, as shown in the photo above. Then connect the feeder coax across the two turn primary and the half-wave wire to one of the secondary wires. Finally, connect a link wire from the other secondary wire to the braid of the coax (see **Fig 2.2**).

A quarter-wave vertical needs an earth or ground plane to work but with an end-fed half-wave there is less current flowing in the ground and it can become almost unnecessary. Despite this, G0KYA recommends using an earth rod and four short radials and also making a choke by coiling about 8 to 10 loops of coax in a six-inch coil about a foot or two from the antenna.

To tune the antenna, use an antenna analyser (or SWR meter with your transceiver) to see where the antenna resonates: it will probably be lower in frequency than 28,000kHz. Cut of a half-inch length of the capacitor coax to reduce the capacitance and

The EFHW matching unit is housed in a waterproof box.

The 20m G0KYA end-fed half-wave during an International Marconi Day portable operation on the Norfolk coast.

move the resonant frequency higher. If you get down to about four inches and are still not there try removing a turn off the secondary coil: it is better to remove turns than to cut off too much of the coax.

G0KYA ended up with 15 turns on the secondary and coax about 4in (10cm) long. The result was an SWR of 1.2:1 at resonance on 28500kHz and an SWR of less than 2:1 across the whole 10m band. To complete the antenna, put the matching unit in a waterproof box.

A version of the antenna was also made for the 20m band. Using a calculator and the equation for the resonant frequency of an LC network:

$$f = \frac{1}{2\pi\sqrt{LC}}$$

it can be seen that halving the frequency means that the value of capacitance must be multiplied by four to make the antenna system resonant. G0KYA

Band	Frequency	Coax capacitor length	Half-wave wire length
10m	28500kHz	13cm (5in)	5m (16ft 5in)
12m	24900kHz	15.25cm (6in)	5.72m (18ft 9in)
15m	21225kHz	19cm (7.5in)	6.72m (22ft 0.2in)
17m	18100kHz	24.5cm (9.5in)	7.87m (25ft 9.8in)
20m	14175kHz	35.5cm (14in)	10.05m (32ft 11.6in)

Table 2.2: Starting lengths for the EFHW matching coax capacitor (trim by 1 to 2in to tune) and half-wave vertical wires (assuming insulated wire is used), for each band from 10m to 20m.

simply cut another length of coax for the capacitor four times as long as the piece used in the 10m EFHW and, with a vertical wire radiator 10.05m (33ft) long, the MFJ antenna analyser showed an SWR of 1.1:1 at 14150kHz, rising to just 1.5:1 at either end of the band.

The recommended 'starting point' lengths for both the coaxial capacitor and the half-wavelength wires for each band from 10m to 20m are given in **Table 2.2**.

Tests showed that the 20m EFHW was at least as good as a half-wave dipole at 30ft and US stations were "romping in" during the afternoon on SSB. It outperforms a regular quarter-wave vertical with radials laying on the ground by a couple of S-points and is a lot easier to put up.

The antenna was also used at a Jamboree on the Air station in Norfolk where it outperformed a G5RV at 30ft by about one S-point.

G0KYA MULTIBAND PORTABLE VERTICAL

Steve Nichols, G0KYA
G0KYA website

After publication of the October 2010 article in RadCom, Steve Nichols provided more details and some interesting updates on his website **[4]**, *including this suggestion for a four-band portable vertical antenna based on a 10m-long vertical wire.*

Steve realised that the impedance at the end of a full-wave wire is similar to that of a half-wave. In other words, if the 10m matching unit described

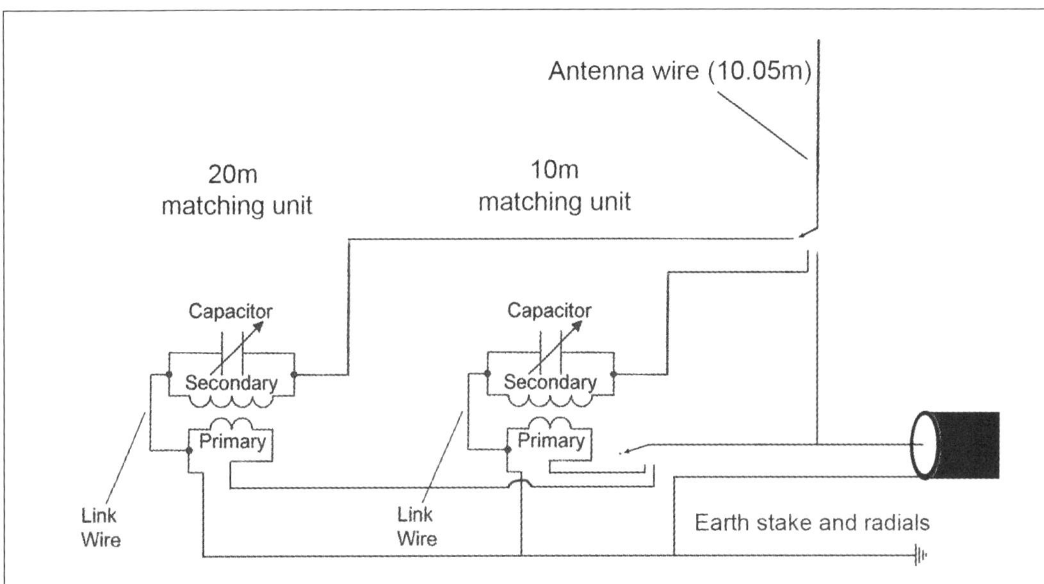

Fig 2.3: Switching between the 20m EFHW matching unit, the 10m matching unit and direct (with a radial system) allows a 10.05m vertical wire to be used on 40, 20, 15 and 10m.

above is attached to a length of wire that is a full-wave long there should still be a reasonably good match. He attached the 10m band matching unit to the bottom of the 20m-band EFHW wire (10.05m long) and found that, sure enough, he had quite a good match at around 28700kHz.

The 10.05m vertical radiator is also a quarter-wave on 40m and three quarter-waves on 15m, which means that you can also feed the wire directly (without any matching unit) and use it as a quarter-wave vertical on 40m, provided you have a good earth such as eight to 16 radials about 10m long. This will also work reasonably well on 15m as a three quarter-wave vertical.

The radiation pattern will not be as good for DX on 15m or 10m because there are some higher-angle lobes when the vertical radiator becomes longer than about 5/8-wave – but it will still work.

Switching between 'direct' (with radials) for 40m and 15m, the 20m EFHW matching unit, and the 10m matching unit (**Fig 2.3**) provides a way of making a 10.05m length of wire work on four different frequency bands. It will be an efficient radiator on 40m and 20m and although something of a compromise on 15m and 10m it should work, albeit with a slightly higher SWR.

A TAPE MEASURE VERTICAL

David L Poole, AD9DP
QST, November 2010

David, AD9DP, describes a temporary portable vertical ground plane antenna that is easy to deploy, resonates well and can be quickly tuned between bands. As with many amateur radio antenna projects in the USA, this one got its start at an ARRL Field Day...

David wrote, "We had limited air space and few trees to hang additional horizontal dipoles, so I wondered how I might build an economical vertical antenna for 20 metres. Knowing of the SteppIR verticals that use an extendible metal strip for a radiator, and of tape measures having been used for ground radials, I wondered if I could use a tape measure supported within a PVC pipe for the vertical element as well." But would this work?

Opinion on PVC in print and on the web is mixed. Some recommended white schedule 40, if anything; others warned of chalk or other additives used by some manufacturers that can degrade RF properties. As for structural strength, some hams reported using PVC for temporary masts up to 20 feet. This seemed reasonable for portable use, considering the availability of other hams to assist with erecting and guying it. PVC pipe and tape measures are as close as the nearest hardware store and available at reasonable cost.

I was concerned about making electrical contact with the blade, so I decided to try that as a first step. A check of DC resistance suggested good electrical contact, particularly after a little run-in around the contact point. I ran some *4NEC2* models to see how effective spring steel would be as radiator and radials. Figuring the RF performance would be no better than the DC resistance would predict, I measured the blade resistance at 0.0245Ω /ft, or 5.94 x 10^7S/m conductivity for use in the models.

Conventional wisdom says that elevating the radials of a ground plane antenna minimises the ground losses and sloping them towards ground raises the radiation resistance. My *4NEC2* models confirmed the desirability of elevated, sloped radials and provided rough physical dimensions for 20m and up with a bare vertical element. *NEC2* has no provisions for modelling a PVC pipe, but tuning would require trial and error anyway.

I chose a 4 foot aluminium stepladder as a stand for the base and also the inner support for the elevated radials. This could as well have been PVC pipe, but a ladder has the advantage of providing a few seconds of self standing ability under calm conditions to allow one person to erect the antenna. Mine has fallen over a few times while I rushed around to guy it, but it fell gracefully and without damage each time.

The rest of the structure was made of PVC pipe and fittings as shown in the photo below. A 1.5in hole cut into the side of the T aids in assembly and allows reading the vertical element length for tuning. One side leg holds a PVC threaded plug with an SO-239 coax socket.

The top leg mounts the first mast section of 1 inch PVC through a threaded bushing. A 1 inch tape measure blade fits within a 1 inch PVC mast by replacing its end hook with a nut and bolt to prevent retraction into the housing.

View of antenna base mounted on a step ladder.

The tape measure housing is held by friction in a slotted pipe in the bottom leg of the T. The remaining side leg can be left open, capped or used for an alternate feed point.

I made a base to attach radials out of 1/2 inch recycled PETE sheet from a local home store. That material has been discontinued, but a polyethylene (PE) or polypropylene (PP) cutting board would provide an alternate source of raw material. I cut an octagonal piece with a centre hole sized for an interference fit and solvent welded it on to the lower leg of the cross T. Four bulldog clips hold the blade ends of the ground radials and connect them to the flange of the SO-239 socket. The assembly mounts to the stepladder with two carriage bolts. I made three 5ft mast sections of 1 inch PVC pipe with male and female couplers solvent welded at each end. A guy ring was fitted over the male threads between the upper two sections.

The first outing was to a local park where local hams meet on Saturdays for coffee, technical talk and to teach occasional licence classes. The complete kit of ladder, feed T, mast sections, tape measures, guy ring, wires, tent stakes and feed line is shown in the photo top right.

We assembled two 5 foot mast sections with the guy ring in between and attached polyester guys. Three hams provided tension on the guys while I coupled the upper two sections to a third then lifted and secured the completed mast into the top of the feed T on the ladder.

The vertical tape measure was stowed and the radial rules stretched out to length, braked and staked to ground (photo right centre). Tape measures with good brakes are helpful for tensioning the radials to minimise droop.

Tuning took but a few tries to find 1.1:1 VSWR at the band centre (bottom right). We were good to go on 20m in about 15 minutes from kit to first QSO. The VSWR was indistinguishably flat across the band. Several hams made contacts, including a number with an Oregon QSO Party, but we had nothing to compare the tape measure vertical with and I had a licence class to teach. At the end of the morning, the tape measure vertical came down just as quickly as it had gone up and it was back in the car.

The tape measure vertical was used on our 20m get-on-the-air station on Field Day. At many times of the day, 40 metres is more useable than 20m, so we wanted to move lower in frequency. Given the mast loading, a 33 foot tape measure could be used on that band as a quarter wave, but it would require more guy wires, more amateurs and a larger clear ground radius to set up. The unused leg of the cross T, however, had 'loading coil' written all over it, so I decided to try the 20m arrangement as a base-loaded vertical on 40m.

Based on 'cookbook' calculations, I made a loading coil by hand on an 8in length of 1.25 inch

The complete kit ready to be set up at the park.

Staked radial detail. The end of the radial does not electrically connect to the stake.

The author making one of the initial tuning adjustments to 20m band centre.

schedule 40 white PVC pipe using one end to become the form and the other as a handle for holding during the wind. I used a length of 1 inch pipe clamped into a vice as a mandrel. I began with a hole and wire loop to anchor the start of the winding 3/8 inch from the form end. I wound 20 turns of #14 AWG bare copper at 10 turns per inch, then secured the far end into another hole with a wire loop. I drilled additional holes around the starting end of the form to anchor trial taps at quarter turn intervals, added a trial tap and headed into the field. With some tweaking, I obtained a match on 40m using 16 turns of the coil and tapping the feed two turns up from the ground end. I later removed the excess turns, secured the winding end and trimmed the form. An alligator clip connects the 16 turn end to the feed rollers for ease of disassembly.

The coil can be left in place on higher bands by disconnecting it from the radials. [If space is available, 40m performance might improve if at least one pair of radials is extended to 33 feet – *Ed.*]

Tests across bands were made with the assistance of Don Backys, K9UQN, and Bill Rowe, AB9SV, using Don's AEA CIA-HF Complex Impedance Analyser. We alternately tuned the vertical and radials for the lowest VSWR at band centre on 10, 12, 15, 17 and 20 metres and on 40 metres with the loading coil. By logging the tape positions on each band, we are able to quickly return to the correct position without test gear.

The 40m performance on a few contacts was reported as comparable reports to an inverted-V at 50 feet. The loading coil remained cool at 790W CW. The optimum dimensions, measured results, and bandwidths to 2:1 VSWR are shown in **Table 2.3**.

It might be said that no amateur antenna project is ever finished, and this one is no exception. The tape measure vertical is a great experimental vehicle. In his *QST* article **[5]**, Rudy Severns, N6LF, observed that four ground level radials perform significantly better when shortened from quarter wave to as little as 0.15 wavelength if the vertical element is lengthened to resonance. While my *NEC2* models suggest there is less of an improvement for shortening 4 foot high sloped radials, shorter radials may be useful if you have a limited area for deployment. Rudy's work also suggests that radial

Band (Metres)	Freq (MHz)	Vertical	Radials	SWR	2:1 SWR Bandwidth (MHz)
10	29.4	85	100	1.3	7
12	25.5	98	105	1.3	5.3
15	21.7	121	125	1.2	4
17	18.1	144	155	1.3	3.4
20	14.2	187	190	1.2	1.5

Table 2.3: Vertical and radial dimensions (in inches) for 10 to 20m.

elevations as small as 1 foot might provide acceptable performance even at 40m. This suggests a stepstool in lieu of a stepladder or even a short section of PVC pipe to elevate the radials.

THE DDRR ANTENNA

Peter Dodd, G3LDO

RadCom 'Antennas' column, February 2014

When is a vertical not a vertical? The answer is when it is a DDRR antenna. The Direct Driven Ring Radiator may not look like a vertical but, as Peter Dodd, G3LDO (SK), explained, that's exactly what it is. The DDRR was invented for military applications in the 1950s and G3LDO described it in his 'Antennas' column in RadCom.

The DDRR is simply a short vertical monopole having a horizontal or flat top extension to provide quarter-wave resonance. In the original design the horizontal section was circular with the end very close to the top of the vertical section (**Fig 2.4**). A capacitor between the end of the horizontal section and the vertical section or ground was used to alter the resonant frequency of the antenna.

Peter Dodd says that "some years ago" he built a rectangular version and tested it as a mobile or portable antenna, as shown in the photo below right. This configuration had the advantage that it can be made up using straight sections of tubing, with the ends joined using 90 degree angle joints.

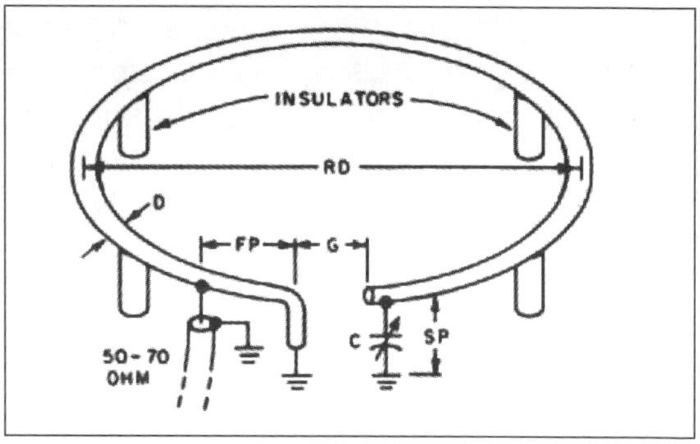

Fig 2.4: The original DDRR antenna.

G3LDO's DDRR antenna feed and earth.

Experimental 20m DDRR 'roof rack' antenna made by G3LDO. Tuning is accomplished by altering the length of the element end near the vehicle roof.

The radiation resistance of this antenna is very low so for the power available with a modern mobile transceiver the RF current is very high. The same high current also exists in the antenna counterpoise, in this case the vehicle body nearest to the antenna. It also means that the connection from the base of the antenna to the metal work of the vehicle must be robust. The antenna was shunt fed by tapping the feed line centre up from the ground end of the element.

The DDRR antenna worked very well: it performed as well as a roof-mounted 8ft (2.5m) loaded mobile whip, although the bandwidth was narrow and required retuning if the operating frequency was altered. Also the RF current flowing in the car bodywork caused some of the car front panel LED indicators to flicker. These tests were performed some years ago on an old vehicle before the days of microprocessor controlled engine management systems. You might not be able to get away with such a system in a modern vehicle, but there was no doubt the antenna worked. It is therefore probably better to consider a car-mounted DDRR as a portable antenna, rather than a mobile one!

G0CBM PEDESTRIAN MOBILE VERTICAL

Peter Dodd, G3LDO

RadCom 'Antennas' column, January 2012

The G0CBM antenna with back-pack.

"Charles Wilkie, G0CBM, a keen HF mobile and pedestrian mobile operator, has come up with an interesting antenna." So wrote Peter Dodd, G3LDO, in his 'Antennas' column in RadCom. He described the design and construction of G0CBM's antenna which overcomes the difficulty of feeding a pedestrian mobile antenna with a counterpoise.

The design is basically a centre-loaded half-wave antenna for 14 and 18MHz, although this frequency range can be extended, as will be described later. The physical size of an antenna can be considerably reduced by using a loading coil and end capacitances, which is done in this case. Traditionally, small antennas are loaded using a solenoid shaped coil. To get the best Q and low loss from a solenoid coil the spacing between each winding of the coil should be at least the diameter of the wire. Feeding such a half-wave antenna with the loading coil at the centre can be a problem.

G0CBM's design uses a disc-shaped loading coil, which he describes as a 'basket weave' coil, although a more accurate description is a 'spider' coil. This type of coil construction is favoured by people who make crystal sets; construction details of the coil used with this antenna are given later.

The antenna is fed via a link-coupling coil. This again is similar to crystal set technology, where coupling is varied by altering the distance between the coupling coil and the loading coil. A further advantage is that it is easier to suppress common-mode currents on the coax feeder, which can cause RF on the case of the radio and 'hot microphone' problems.

The antenna support comprises a 2.4m telescopic fishing rod with a capacitance hat of six 0.3m spokes fixed to the top. The vertical wire element is simply a length of insulated wire loosely wound around the fishing rod from the inner part of the loading coil to the capacitance hat.

The disc-shaped loading coil is fixed one metre from the bottom antenna support, with the adjustable coupling coil located just below.

Four 0.6m long radials are fixed 0.9m from the lower end of the antenna support; this lower end acting as a 'mounting pole'.

The complete antenna weighs only 950g.

The antenna support is constructed from a fibreglass telescopic fishing rod with the top two or three sections removed to ensure that it stays rigid enough to withstand the whipping motion caused by walking. The fishing rod needs to be of good quality with tight fitting sections.

The six capacitance hat spokes at the top of the antenna are made from 1.6mm stainless 'tungsten inert gas' (TIG) welding wire pressed into a brass hub, which is then screwed on to a brass bush at the top of the fishing rod.

The inner sections of the four radials are each made from to 0.3m copper plated steel tube (salvaged sections from a military manpack whip), fitted to the brass centre boss. The ends of the radials are made from 0.3m of 1.6mm stainless TIG welding wire.

G0CBM has the necessary facilities for turning the capacitance and radial bosses and the loading coil sliding bush, although he has used salvaged CB antenna parts in the past. TIG welding wire was used for the capacitor hat and the outer part of the radials because it can be easily reformed if it is bent after snagging on bushes or trees.

The loading coil former is made from a 90mm diameter disc of 6mm polycarbonate, although thick PCB (without copper) could also be used. Nine radial slots are cut into the polycarbonate using a 5mm cutting disc and angle grinder. The odd number of slots in the disc allows the wire to lay alternatively from one side to the other on each successive turn, creating a space between each winding of the coil. This effectively gives an approximate adjacent wire spacing of the diameter of the wire, which reduces the self capacitance of the coil and improves the Q. The coil comprises 12 turns of 2mm enamelled copper wire wound into the slots. The loading coil is shown in the photograph to the right. A spade connector is soldered to the inner end of the coil to connect the wire from the capacitor hat.

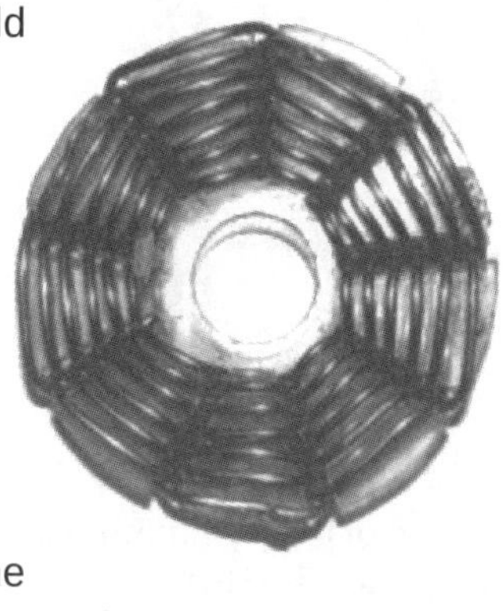

The centre loading coil.

The coupling coil.

Tapping points are soldered on to the last turn of wire at every segment to connect the flying lead from the radials.

The coupling coil is made from a 75mm disc of PCB. Three spiral turns are etched on to the copper laminate and the coax feeder soldered to the ends of the coil as shown in the photo to the left. The copper laminate coil is flow soldered to reduce tarnishing and probably improve efficiency. The coupling coil is fixed on a sliding bush just below the loading coil and connected to the transceiver via a short length of 50Ω coax. Ferrite sleeves are threaded on to the coax near to the coil to reduce common-mode currents. A flying lead is used to connect radials to the any one of the tapping points on the outer part of the loading coil, as shown in the photograph below.

An antenna analyser such as the MFJ-269 or a noise bridge is connected to the short coax lead on the coupling coil. The required resonant point is set by connecting the flying lead from the radials to the appropriate tapping point on the loading coil, indicated by a minimum SWR or the nearest you can get to 50Ω impedance. (If you have a GDO this can be used to set up resonance without any connection to the coupling coil.)

Adjustment for the lowest SWR is achieved by sliding the coupling coil closer to or further from the loading coil. The resonant frequency will change slightly as the coupling coil coupling is adjusted, so final adjustment to the loading coil taps might be necessary.

It is important that a very low SWR is obtained at the frequency of interest. This ensures maximum efficiency and the widest low SWR bandwidth. The position of the taps and the coupling coil can be marked for rapid reset when the frequency is changed.

The antenna works well on 17m by adding more taps to the third turn from the outside of the loading coil and clipping the radial flying lead to any one of these. The coupling coil is adjusted for minimum SWR in the same manner as for 20m.

Further taps can be added for 15 and 10m but a significant proportion of the loading coil will be left open, which may result is a lower efficiency. The solution may be to make an additional loading coil with fewer turns plus a coupling coil with just two turns for these higher frequencies.

Some mobile antennas such as the 'Texas BugCatcher' short out the unused turns with the flying lead rather than leaving them open. When I asked G0CBM if he had considered this approach he replied that experimental work done with G0UOO indicated 'free' or 'bridged out' turns reduce efficiency.

Details of the positioning and connections of the loading coil, coupling coil and radials.

The various components that make up the antenna are shown in the photo. The TIG wire extensions of the radials are folded back along the steel tubing sections.

I am not particularly keen on anecdotal descriptions of antenna performance but in this case G0CBM is using QRP power with a very small antenna. To work any DX at all with such a pedestrian mobile installation means that the antenna is certainly no dummy load. G0CBM goes on to say, "The performance on the 17 and 20m bands is very good and the majority of stations called have replied. With just 5W from an FT-817, I had a QSO with an American serviceman YI9GYS, in Tikrit, Iraq, from the beach at Skegness. While on holiday in the Algarve, Portugal, again using 5W, I had a QSO with Ron, VK3IO, in Melbourne." G0CBM has been a keen pedestrian mobile operator for many years and makes the following comments about it. "Although pedestrian mobile may seem a disadvantage at first because of the small antenna and 5W, full advantage can be made of the effects of sea and summits to aid propagation. The suffix 'pedestrian mobile' always gives at least a 10dB advantage over the competition as well."

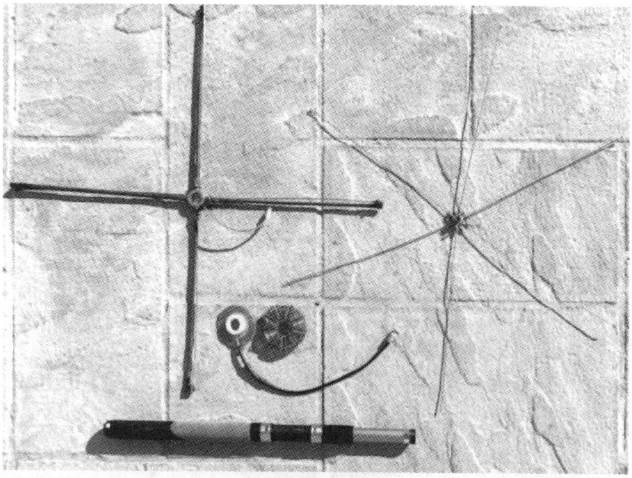

The antenna components.

AN EASY TO MAKE TWO BAND HF GROUND PLANE

Joel Hallas, W1ZR
QST, December 2013

The tree-branch supported vertical ground plane is a popular antenna for portable operations, but it is usually a single-band affair. Here's a clever way of making one antenna work on two bands. The design is based on a two-band coupled resonator dipole made from window line. That antenna works very well, is easy to make and easy to duplicate. Joel, W1ZR, said that he had received a request to adapt it to vertical polarisation.

W1ZR writes: "The original design could, of course, be adapted for use as a vertical dipole, and some have successfully done so, particularly for VHF FM operation. It turns out that a coupled resonator can also be configured as a vertical ground plane in order to minimise the height of the required support. This makes sense, since a vertical dipole needs a support more than half a wavelength high. With a single available support at that height, for most HF applications, an inverted-V dipole provides improved performance."

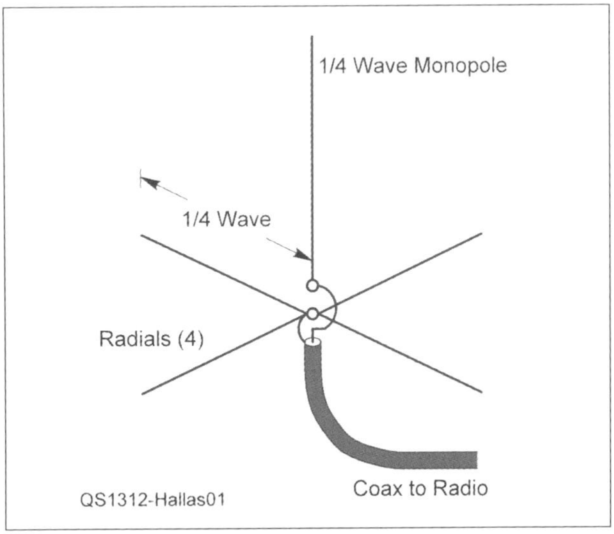

Fig 2.5: A traditional ground plane antenna. The four horizontal rods (radials) serve as an artificial ground for the monopole. Note that while they serve in place of a ground, they are not at ground potential. The ends of the radials are at the same potential as the tip of the monopole, so people and animals should be protected accordingly.

A ground plane antenna is a vertical 1/4 wave monopole mounted and fed against an artificial ground system – the 'ground plane'. As originally developed, the artificial ground consisted of a metallic disc with a radius of 1/4 wavelength or more. While that is feasible at UHF, at lower frequencies, the ground plane is usually constructed of multiple 1/4-wave conductors, called radials, perpendicular to the vertical element and located at its base, as shown in **Fig 2.5**. Configured in that way, the antenna has a feed point impedance that is half that of a dipole, or 36Ω, in free space.

While the configuration of **Fig 2.5** is often encountered – sometimes using a gamma match to provide a feed impedance of 50Ω, another alternative is to lower the outboard end of the radials to make a drooping ground plane. This skeleton conical arrangement is topologically somewhere between the standard ground plane and a vertical conical dipole.

Such an arrangement can provide a good match to 50Ω coax. We have taken that approach here, although to use even less vertical space, if a small mismatch can be tolerated on the lower frequency band, horizontal radials can be used.

As with the two-band horizontal dipole, this antenna is really two parallel ground planes made from window line. I selected the 20m and 15m bands for this model, however, it could easily be adapted to other pairs of bands. The lower frequency ground plane is fed directly with 50Ω coax, while no portion of the higher frequency ground plane is connected directly to the

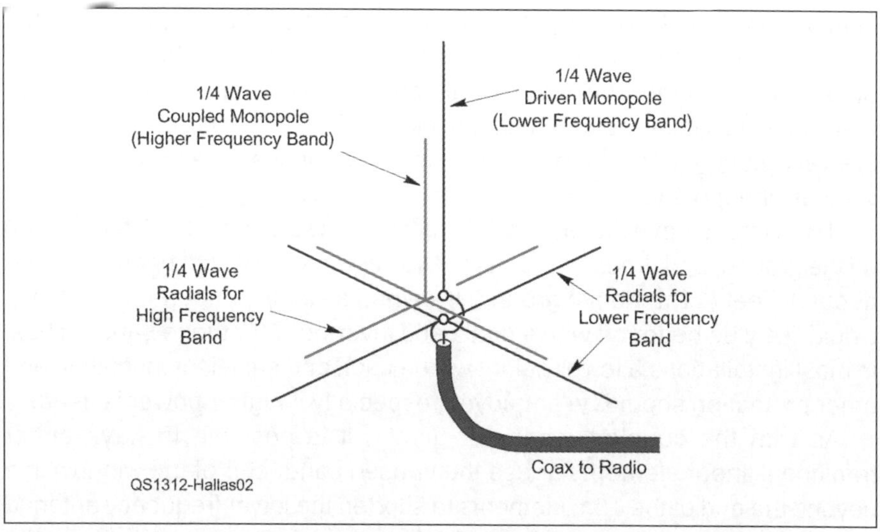

Fig 2.6: Simplified drawing of the two band coupled resonator ground plane. Note that there are no electrical connections to the higher-frequency elements.

driven unit, but receives its energy through parasitic coupling (see **Fig 2.6**). While three, four, or even more radials could be used, the two radial configuration I used is within less than 1dB of being omnidirectional and works very well for those with physical constraints. Some have even used a single radial – picture a bent vertical dipole configuration – but that is more directional and has a higher angled main lobe, so we used two here.

A vertical monopole can be ground mounted, but works somewhat better if off the ground. Another consideration is that the ends of the radials are a

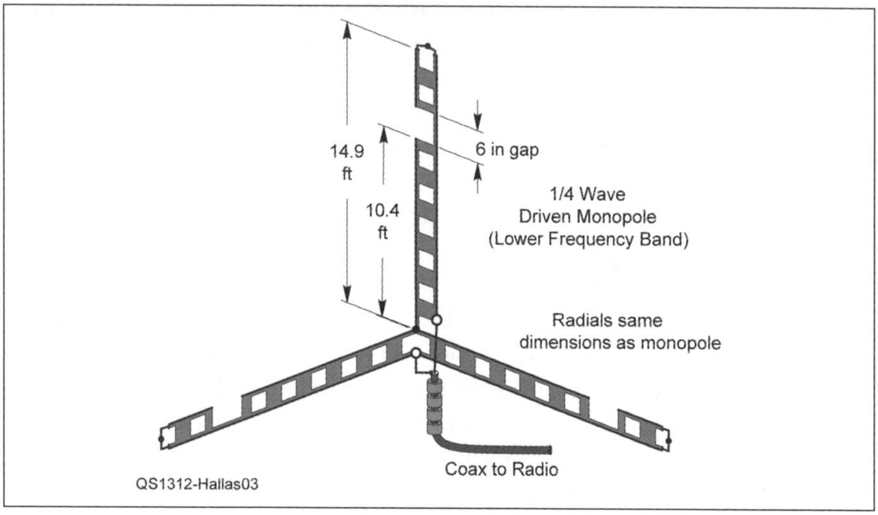

Fig 2.7: Dimensions of the window line 20 and 15m ground plane. For best match the radials should droop from the horizontal. While 45° is optimum, it will work satisfactorily with a less extreme droop.

high voltage point, with a potential comparable to that at the top of the monopole. Thus for optimum safety, those ends should be high enough or otherwise protected from accidental contact. It is important to remember that while the radials serve as an artificial ground, in this configuration they are very much a part of the antenna and must be insulated as well as the vertical monopole.

If we were to have the ends of our radials at a safe height of 8 feet (2.4m) off the ground, and droop our radials at 45 degrees, we would need a support about 36 feet (11m) above ground. If we had a single support that high, we would likely be better off with a dual band inverted-V for these bands. Thus, in most installations a lower height will be used and sufficient insulation and other protection should be employed, especially if higher power is used.

As with the coupled resonator dipole, it is possible to have either traditional linear elements or use the unused conductor of the window line beyond the end of the 15m elements to shorten the lower frequency antenna elements in a 'folded' arrangement. At first, I was just going to describe the linear arrangement, but then decided that since available height was a key issue, I would save a bit more than a foot by using the folded approach. There really is no downside to this arrangement. Note that in the folded configuration, the highest voltage points are at the gap.

The exact dimensions will depend on the height above ground and the ground conditions. The dimensions shown in **Fig 2.7** were set with the base 12 feet (3.66m) above ground. I lowered the antenna so that the feed was at 3 feet, and the radial ends were at about 1.5 feet. As a result, the 20m resonance dropped by 270kHz, while the 15m resonance dropped by only 20kHz. Of course, this is a good direction in which to have a change, since it's easier to trim than 'stretch' antenna wire. My suggestion would be to start with a few inches more than shown and trim as needed. By the way, order a few more feet of line than you think you will need – my order came up a few feet short.

Close-up of the feed point arrangement showing the six ferrite beads below the feed point that make up the common mode choke.

The window line uses #18 AWG stranded copper covered steel wire. While other types of window line could be used, different wire size or dielectric properties will have an effect on the final dimensions, so plan accordingly. Note that while it looks like there are three distinct pieces, the two radials can be made from a single length of window line with the insulation stripped at the centre of each piece.

I used a common mode choke made from six slip-on ferrite beads taped in place just below the feed (see photo to the left). Of course, other kinds of chokes could be used, however, the beads noted slip nicely over RG-8X, don't require tight turns as on a toroid and are lightweight and compact.

While the design shows the monopole and radials the same length, if you are close to being on frequency, you

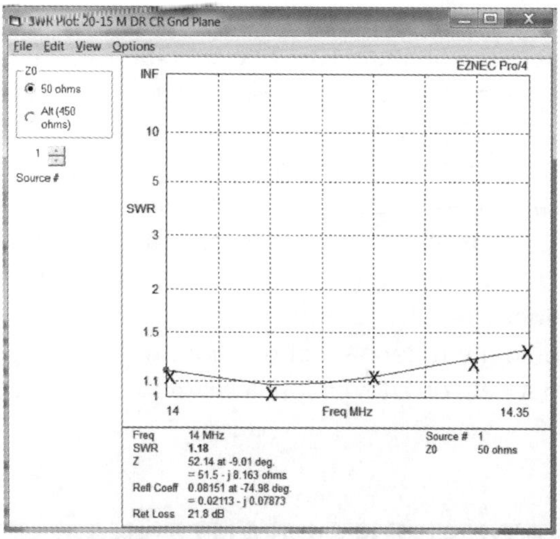

Fig 2.8: *EZNEC* predicted SWR of the folded coupled resonator ground plane on 20m. The measured results taken through 25 feet of RG-8X are shown with the Xs.

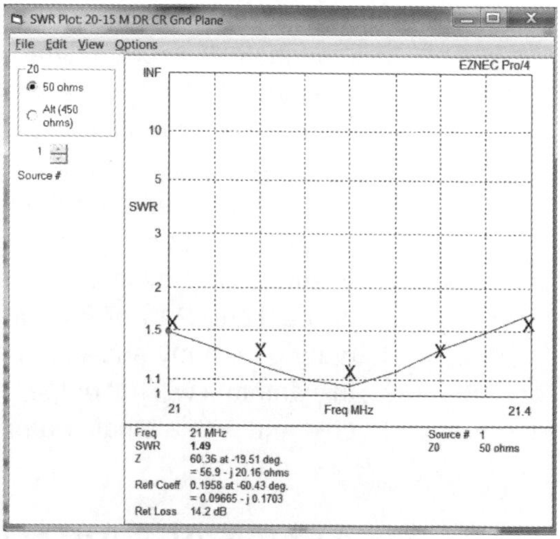

Fig 2.9: Same as Fig 2.8, but for 15m.

can just trim the monopole. It makes the antenna the ground plane equivalent of an off-centre fed dipole. Any unbalance resulting from the small difference will be largely eliminated by the common mode choke.

The antenna came together quickly and works just as *EZNEC* **[6]** predicted. The measured SWR for both bands at the end of 25 feet (7.6m) of RG-8X is shown on top of the *EZNEC* SWR predictions in **Figs 2.8** and **2.9**, above, while **Figs 2.10** and **2.11** (below) show the elevation patterns on both

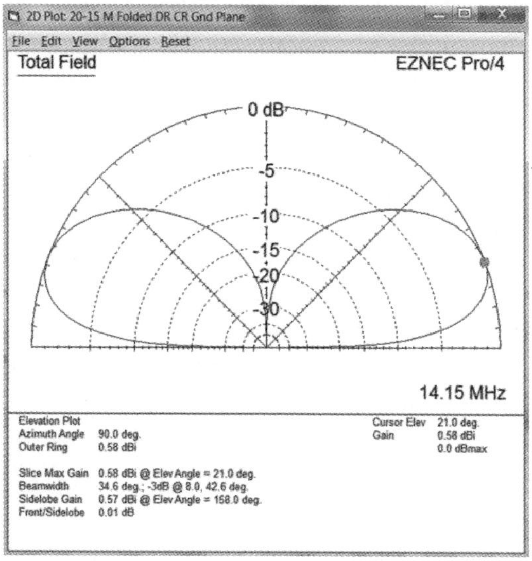

Fig 2.10: *EZNEC* predicted elevation pattern on 20m.

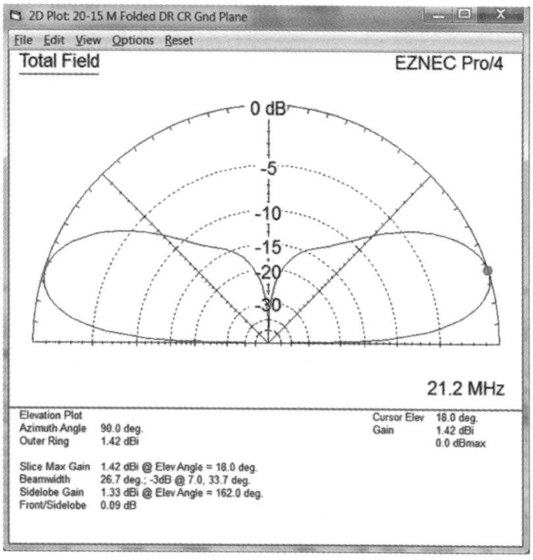

Fig 2.11: Same as Fig 2.10, but for 15m.

bands, with the antenna feed 12 feet (3.65m) above typical ground (conductivity 0.0005 S/m, dielectric constant 13). The unfolded version (16.25ft / 4.95m high) at the same feed point height has a gain only 0.1dB higher. The azimuth pattern is within 1dB of being omnidirectional.

Tests on the air tended to confirm the predictions. In A/B comparison with a three-element triband Yagi at about 35 feet (10.7m) high, the Yagi always did better by around 3 S-units for stations at medium range that needed high elevation angles. DX stations were usually within 1 or 2 S-units of the Yagi. Part of the difference was due to the extra 2dB of line loss getting from my entrance panel to the antenna test position. In summary, this antenna will not outperform a Yagi on a tower, but it will work DX at low cost and with a small footprint.

A COAXIAL CABLE VERTICAL DIPOLE ANTENNA

Al Yerger, K2ATY
QST, May 2014

Here's another ingenious idea for getting on the air quickly from a portable location when a tree branch is the only method of supporting an antenna.

Vertical antennas have long been utilised for communications throughout the spectrum for any number of reasons. An ideal vertical antenna has a low angle of radiation, keeping most of the transmitted energy close to the horizon, providing excellent long distance propagation. In addition, they provide omnidirectional radiation patterns and can be useful for portable operating and / or in limited space situations.

Vertical antennas also have some disadvantages. Among them is the requirement for a rather robust ground system when the antenna is ground mounted, or a system of elevated quarter-wavelength radials if the antenna is mounted above the ground. Another problem with a traditional quarter-wavelength vertical is that most of the radiation from a vertical antenna takes place near the bottom of the antenna where the antenna current is highest. Unless the antenna is mounted high and in the clear with the requisite radials, most of your signal is subjected to various forms of ground clutter, such as nearby buildings. Also if your ground plane is not perfect, the maximum radiation will not be at the low angle you want and expect. In an attempt to raise the point of maximum radiation and solve some of these problems, end-fed half-wavelength verticals are sometimes used, but they present high feed point impedances along with other matching issues.

Another approach to a vertical half-wavelength antenna is the coaxial vertical. A coaxial vertical is simply a half-wavelength dipole (**Fig 2.12**)

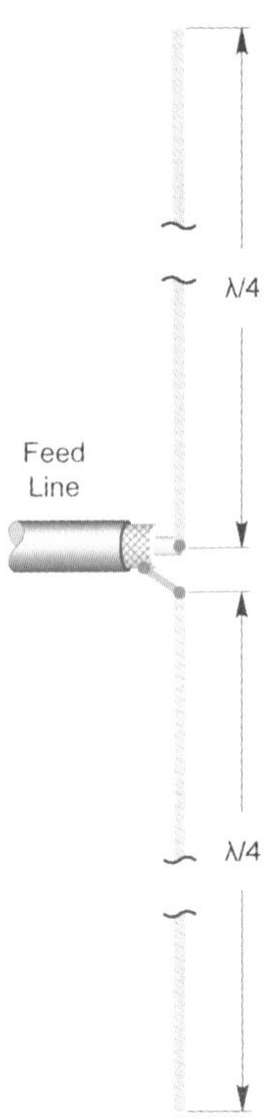

Fig 2.12: The vertical half-wave dipole. Because it is a dipole, a ground radial network is not required, but care must be taken to keep the feed perpendicular to the dipole to prevent coupling on to the feed line.

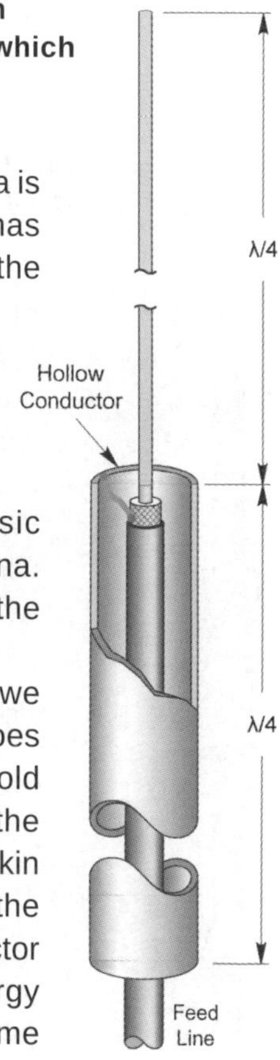

Fig 2.13: The coaxial vertical dipole. 'Coaxial' refers to the tubular bottom element, which serves to route the feed line to the centre of the dipole, which avoids feed line coupling.

orientated vertically. The lower quarter-wavelength portion of the antenna is hollow and the transmission line is run inside the radiator. This design has been used for years in VHF and UHF antenna design. **Fig 2.13** shows the construction of a basic half-wavelength coaxial vertical.

For HF antennas this design is not always practical due to the size of the antenna. For bands below 10m, a coaxial vertical of conventional design may be simply too heavy to properly support. The design we are presenting here combines the properties of a line isolator, acting as a common-mode choke, along with some basic transmission line theory to create a practical HF coaxial vertical antenna. The lower frequency limit of the design will be dictated by the height of the available support.

To start the discussion regarding the theory behind this antenna, we need to ask a question that I am sure everyone has heard somewhere; does current flow *on* or *in* the shield of the coax? The answer to that age old question is 'yes' and 'no'. The reason for the ambiguous answer lies in the operation of coaxial transmission line and a phenomenon known as the skin effect. Skin effect is the principle that says RF energy tends to flow on the surface of a conductor. The amount the energy penetrates into the conductor depends upon frequency. The higher the frequency, the less the energy penetrates into the conductor. From a practical standpoint, we can assume that all of the RF energy flows on the surface.

Now let's take a look at coaxial cable. **Fig 2.14** shows the construction of typical RG-213 type coaxial cable. There is a dielectric and then an inner conductor surrounded by an outer conductor or shield. Applying the skin effect principle, we see that current flows on the outer surface of the inner conductor and on the inner surface of the outer conductor. If the cable is properly terminated at each end, there should be no current flowing on the outer surface of the shield. The primary purpose of a balun or line isolator is to ensure that RF current does not flow on the outer surface of the transmission line.

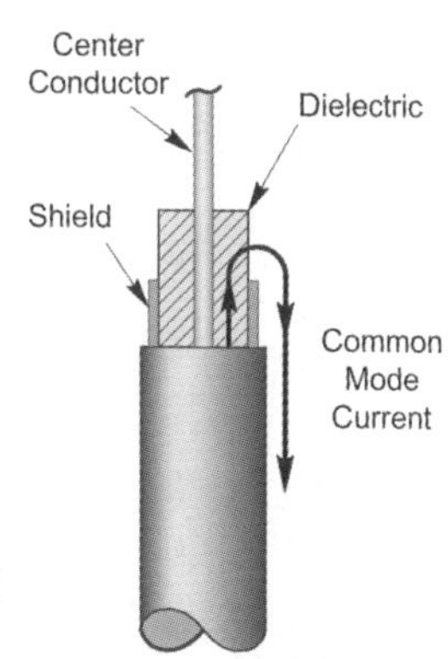

Fig 2.14: Construction of typical coaxial cable such as RG-213 or RG-8X.

For this antenna, treat the inner and outer surfaces of the coaxial cable shield as if they were separate conductors. The outer surface of the shield becomes the sleeve portion of the coaxial vertical. To obtain resonance, the sleeve must end at the proper length to create a half-wavelength dipole. To set the length of the sleeve, utilise a line isolator. I used a Radio Works T4 line isolator, but any high quality line isolator should work.

The line isolator places a very high inductive reactance in series with the outer surface of the cable creating essentially an open circuit. The location

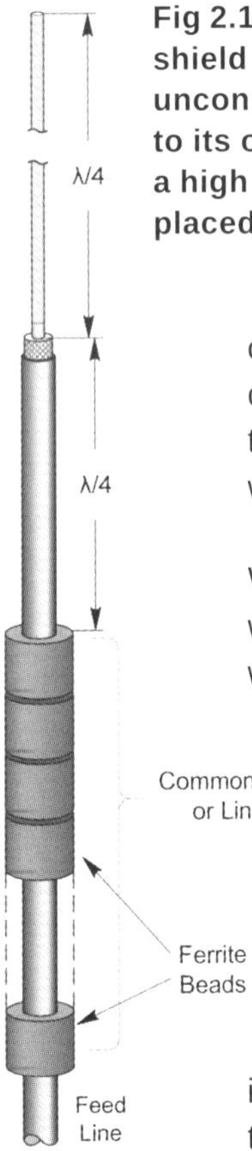

Fig 2.15: The coaxial cable vertical dipole that uses a portion of the feed line shield to serve as the bottom element of a vertical dipole. The feed line shield is unconnected at the centre of the dipole so current flows from the shield's inside to its outside, where it flows down as a common mode current until it encounters a high impedance presented by a common mode choke, acting as a line isolator, placed a quarter-wavelength away.

of this open circuit establishes the end of the lower element half of the coaxial vertical. To create the upper vertical element, we simply extended the centre conductor with a quarter-wavelength piece of #14 AWG copper wire.

The current inside the coax flows up through the line isolator unaffected. When the current on the inside of the shield reaches the top of the coax it will then flow down the outside of the shield until it reaches the line isolator where, due to the inductance of the line isolator, the shield will appear as an open circuit effectively creating the lower end of the dipole. Due to skin effect, the two currents never 'see' each other. This is how we cause the shield to function as if it were two separate conductors.

See **Fig 2.15**. To construct the lower half of the radiating element, start with a piece of RG-213 type cable. Cut the cable to the length shown in **Table 2.4** for the desired frequency. This length will be about 13% shorter than a quarter-wavelength in free space. This takes into account the fact that the diameter of the cable creates a somewhat thicker element than a single thin piece of wire. This lowers the resonant frequency of the antenna. The table values include a multiplier of 0.96 for the insulation on the cable.

Frequency (MHz)	Lower Element (ft in)	Upper Element (ft in)
3.750	58' 9"	62' 3"
7.150	30' 6"	32' 6"
10.125	21' 5"	22' 11"
14.175	15' 2"	16' 4"
18.115	11' 10"	12' 10"
21.225	10' 1"	11' 0"
24.950	8' 7"	9' 4"
28.300	7' 6"	8' 3"

Table 2.4: Coaxial cable vertical dipole element lengths (dimensions in feet and inches as in the original article. To convert feet to cm multiply by 30.48; to convert inches to cm multiply by 2.54.)

Also accounted for is the length of the connectors between the upper and lower elements. The top of the lower section really ends at the top of the PL-259 that connects the top section to the lower section of the antenna. This is where the current on the inside of the shield actually spills out, on to the outside of the shield. This transition adds about 1.875 inches (about 4.8cm) to the overall length of the bottom section. In other words, the lengths presented in **Table 2.4** are the *cut* lengths of the cables before installing the connectors.

Prepare the lower section by cutting the proper length of RG-213 cable and installing a PL-259 connector on each end of the cable. Be careful that when you prepare the ends of the cable, you only expose the minimum amount of

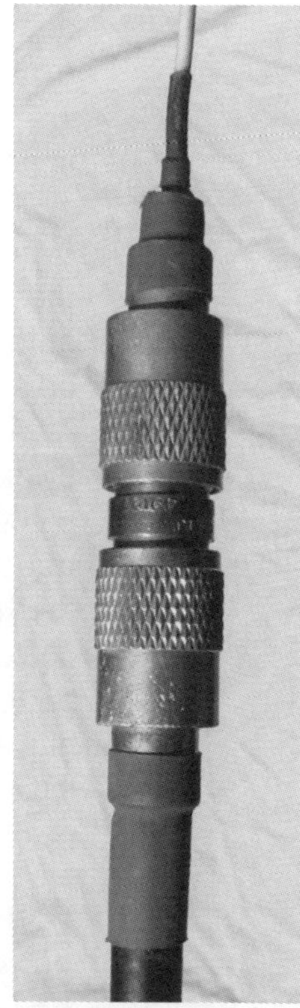

The top element, comprised of #14 AWG wire connected to a PL-259 plug, is joined to the bottom coaxial element through a PL-258 double female connector.

centre conductor and shield required to properly attach the PL-259. Cutting away too much will raise the resonant frequency of the finished antenna. Adjusting the length of the lower element is difficult once the connectors are installed.

For the upper element, measure the proper length of #14 AWG wire and solder it to the centre pin of a PL-259 type UHF plug. Again, **Table 2.4** contains suggested lengths for the various bands. I placed a small piece of the RG-213 dielectric around the bottom of the #14 AWG wire to help insulate the upper wire from the body of the PL-259. I then sealed the top of the connector with heat shrink tubing.

Connect one end of the lower cable assembly to the line isolator, join the upper and lower sections together with a barrel connector as shown in the photo to the right and then hang the antenna from a convenient support using a standard antenna end insulator. Feed the lower end of the line isolator with any convenient length of transmission line.

While the line isolator or common mode choke can provide a high impedance to shield current, it is important to recognise that the end of a half-wave antenna has a high electric field intensity that can couple around the choke and result in strange effects, including unexpected changes to tuning. Adding an additional choke some distance below the primary isolator can reduce this effect.

The antenna can be fine tuned to frequency by simply adjusting the length of the upper element. If you are careful when you construct the antenna, you shouldn't have to adjust the length very much to get a good match across the desired band. If you adjust the length of the upper element too much, you run the risk of unbalancing the antenna. Unless you get to an extreme condition, a little unbalance is harmless. The biggest effect you will probably notice is an increase in the SWR at the resonant frequency. As the feed point of a dipole antenna moves away from the centre, the impedance of the antenna increases, increasing the SWR even though the antenna is actually resonant. The impedance of this antenna at resonance is probably not going to be exactly 50Ω. The closest I could get to 50Ω on my 17m prototype

A typical installation of the coaxial cable vertical dipole hanging from a tree limb. The common mode current choke function is served by a Radio Works T-4 line isolator.

antenna was 69Ω, which is about 1.4:1 SWR. You may be able to obtain an exact match by experimenting with a coaxial matching section; however 1.4:1 still represents a decent match and is actually close to the theoretical impedance of a half-wavelength dipole which is closer to 70Ω than 50Ω.

The antenna is ideally suited for portable operation. Just throw a rope over a suitable support, pull the antenna up and start operating. Because the antenna is a vertical radiator, it would be best to avoid hanging the antenna close to a tower or other metallic support. The antenna could also be suspended inside a fibreglass or PVC pipe.

A possible variation of the antenna would be to build your own line isolator by placing about 12 inches (31cm) of Amidon FB-77-1024 ferrite beads on a piece of RG-213 that would serve as both the lower section and the transmission line (make sure that the beads stay together). Simply slide the beads to the proper position to set the end of the lower part of the coaxial dipole and install the proper top section. This would probably work better on the higher frequencies than it would down at 3.5 or 7MHz, as it is harder to obtain the necessary inductance to choke off the RF current at the lower frequencies.

G3KHZ IOTA DXPEDITION VERTICALS

Steve Telenius-Lowe, PJ4DX
RSGB IOTA Directory, 2014

Derek, G3KHZ, and Mike, K9AJ, set up one of the verticals used on Starbuck Island, T32SI, in 2010.

Often DXpeditions (and in particular DXpeditions for the Islands On The Air – IOTA – programme ***[6]****) take place from locations very close to the ocean. If vertical antennas are used, the close proximity to the ocean gives the signals from these DXpeditions a great boost. Derek Cox, G3KHZ, used EZNEC* ***[7]*** *to model a series of seven vertical antennas for the 10 – 40m bands before building them. The resulting antennas were described in the article 'Antennas for IOTA DXpeditions' in the 2014* IOTA Directory.

It is well known that HF vertical antennas provide a lower angle of radiation than do horizontally-polarised antennas, unless the horizontal antenna is very high above ground. Having such a high horizontal antenna is usually impossible, particularly during a short-duration operation such as a DXpedition.

The sea provides an almost perfect ground plane for verticals and, because it extends many wavelengths from the antenna, losses in the near field (which severely attenuate the very low angle radiation required for long-haul contacts) are much reduced when a vertical antenna is mounted close to the sea. From the above, it is clear that IOTA DXpeditions that are located very close to the sea should always use vertical antennas. How close is "very close"? Unfortunately, the additional 'free' low-angle gain does reduce as you move away from the ocean and more than one or two wavelengths away it has gone. That means on 28MHz your antenna needs to be not more than about 20 metres (65ft) from the sea (and preferably closer), whereas on 160m you could get away with a much greater distance. The best advice is simply to site your antennas as close as is reasonably possible to the ocean. There is another obvious advantage of placing your antennas very close to the ocean and that is that there can be no rising ground or other obstructions such as houses or trees to impair the signal (this applies to both horizontal and vertical antennas, but the additional 'free' gain associated with operating from the seaside only occurs with verticals).

The G3KHZ IOTA DXpedition verticals were designed by Derek, G3KHZ, specifically for use where the antennas would be mounted within a few metres of the ocean. They have been used successfully at ocean-front locations on DXpeditions to many Papua New Guinea islands as well as on four islands in the Southern Line Islands group in East Kiribati in 2010 (T32MI, T32CI, T32VI and T32SI).

The antennas were designed to fit on to Spiderbeam **[8]** 12m long fibre-glass poles. As a result, full-length vertical half-wave dipoles were used on 10, 12, 15 and 17m, loaded vertical dipoles on 20 and 30m, and a loaded quarter-wave ground plane on 40m (no antennas were designed or made for 80 or 160m).

In the case of the four full-size vertical dipoles (10, 12, 15 and 17m), G3KHZ used *EZNEC* to calculate both the length of the antenna and the height of the feedpoint above ground that gave the maximum gain over the sea, together with a 73Ω feed impedance. The calculated feedpoint height of the 20m vertical dipole was 8.05m, which is why a loaded dipole needed to be used on a 12m high mast.

The dimensions and optimum height of the feedpoint are given in **Table 2.5**. Impedance transformers, made up of λ/12 lengths of 75Ω and 50Ω coax, are used to provide a good match to 50Ω feeder (see **Fig 2.16**).

Band	Length of half element	Feedpoint height above ground	Feedpoint impedance
10m	2.59m	4.00m	73Ω
12m	2.94m	4.46m	73Ω
15m	3.40m	5.87m	73Ω
17m	3.94m	6.40m	73Ω

Table 2.5: G3KHZ vertical dipoles for 10 – 17m.

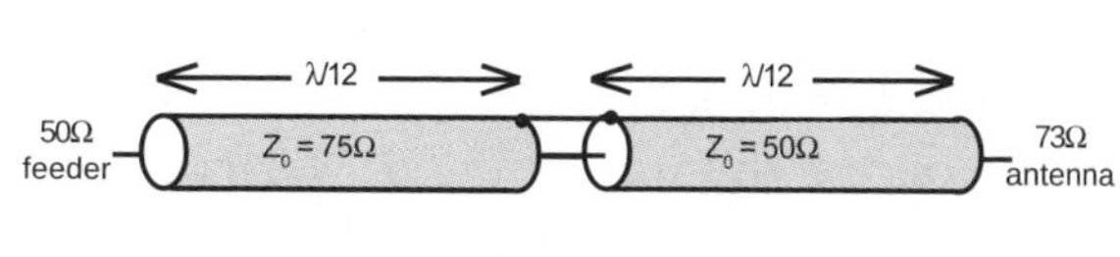

Fig 2.16: Impedance matching transformer for G3KHZ 10, 12, 15 and 17m vertical dipoles.

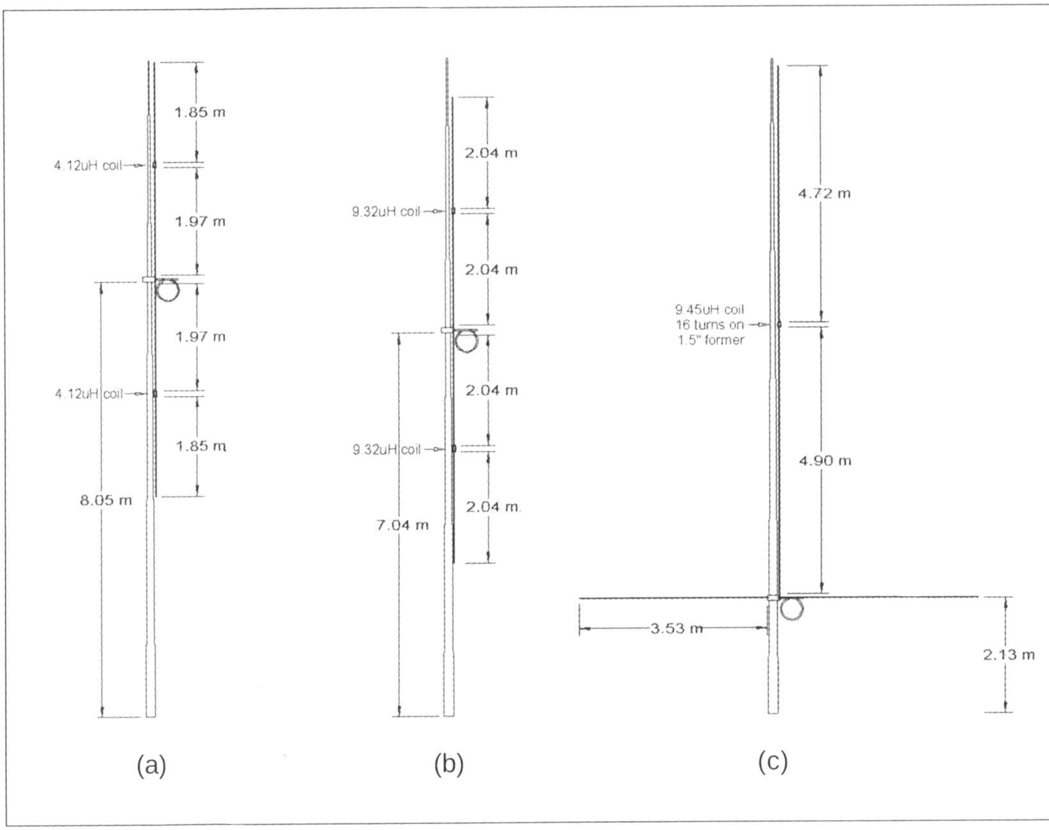

Fig 2.17: Constructional details for G3KHZ IOTA DXpedition verticals for (a) 20m, (b) 30m, and (c) 40m.

The velocity factor of the cables must be taken into account, e.g. for a design frequency of 18.1MHz, $\lambda = 300/18.1 = 16.57$m, so $\lambda/12 = 1.38$m. If the velocity factor is 0.66, the length of the cable is 1.38m x 0.66 = 0.91m. The two lengths of cable are joined using BNC or PL-259 / SO-239 plugs and sockets and are coiled up to form a choke balun.

Band	Antenna type	Gain @ elevation
10m	Vertical dipole	5.73dBi @ 7°
12m	Vertical dipole	6.50dBi @ 7°
15m	Vertical dipole	6.68dBi @ 6°
17m	Vertical dipole	6.69dBi @ 6°
20m	Loaded dipole	6.66dBi @ 6°
30m	Loaded dipole	5.73dBi @ 7°
40m	Loaded ground plane	4.70dBi @ 9°

Table 2.6: Maximum gain (in the direction of the sea) and elevation angle of maximum gain of the G3KHZ IOTA DXpedition verticals.

On the 20, 30 and 40m bands, the 12m poles are not long enough for full size dipoles, so the 20m and 30m antennas are loaded dipoles and the 40m antenna is a loaded ground plane. See **Fig 2.17** for constructional details. These three antennas all have a 50Ω feed impedance so impedance matching transformers are not required on these bands, although it is still important to have a 1:1

A series of the G3KHZ IOTA DXpedition antennas lined up along the sea front at Flint Island, T32VI, in April 2010.

current choke balun at the feedpoint (this is especially the case for the 40m ground plane).

Using the dimensions given in **Table 2.5** and in **Fig 2.17**, *EZNEC* calculates the gain (in the direction of the sea) for each antenna to be as shown in **Table 2.6**. A typical *EZNEC* plot (in this case for the 17m antenna) is shown in **Fig 2.18**.

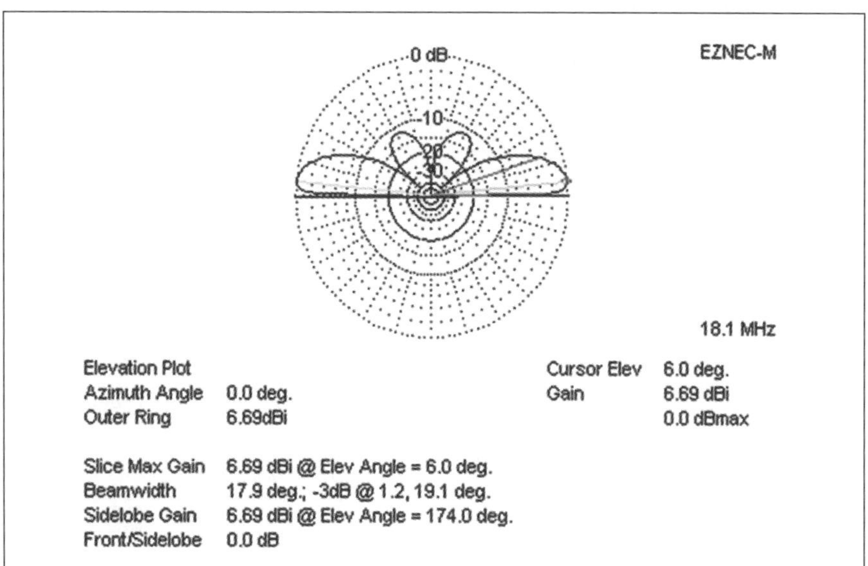

Fig 2.18: *EZNEC* plot of G3KHZ 17m vertical dipole (located close to the sea).

KITE-SUPPORTED VERTICALS

Richard Newstead, G3CWI

RadCom 'Portable' column, November 2008

We conclude this chapter with what is, in fact, the earliest form of antenna: after all Guglielmo Marconi himself used box kites to support his vertical antennas. Richard Newstead, G3CWI, as well as founding the SOTABEAMS company, is a keen portable operator and back in the 1980s used kites to support antennas while working in Antarctica as VP8ANT ***[9]****. In the* RadCom *'Portable' column Richard wrote about his experience using parafoil kites to support vertical antennas.*

Invented by Domina Jalbert in 1964, parafoils have no rigid parts, are easy to fly (even in light winds), stable in fight and fly at a high angle. Look for one with a single line and with an area of between 0.75 and 1.5m^2. Don't be tempted to use anything much larger as it will be difficult to control in a medium to strong breeze. Parafoils are readily available in kite shops.

Most people will fly either a sloping wire or a vertical. There are two main ways to do this. The first and most obvious is to use the kite line itself as the antenna. Use a header of at least 15m of normal kite cord before attaching the wire; this will help you in launching the kite. Use a light stranded wire. **Fig 2.19(a)** shows the general arrangement. Note the use of a post at the ground end to give strain relief for the connection to the radio equipment – it is inconvenient to have to chase your radio across a field. Stakes sold for dogs are particularly suitable here. Another tip is to include a connector in the antenna near the radio that will easily pull apart, such as a banana plug and socket. This will also help stop the radio being dragged away should something go wrong.

The second method is to drop a vertical down from the kite line. Again use a substantial header (at least 15m) as this will help the kite fly better and keep the antenna more stable. Although this method can result in a

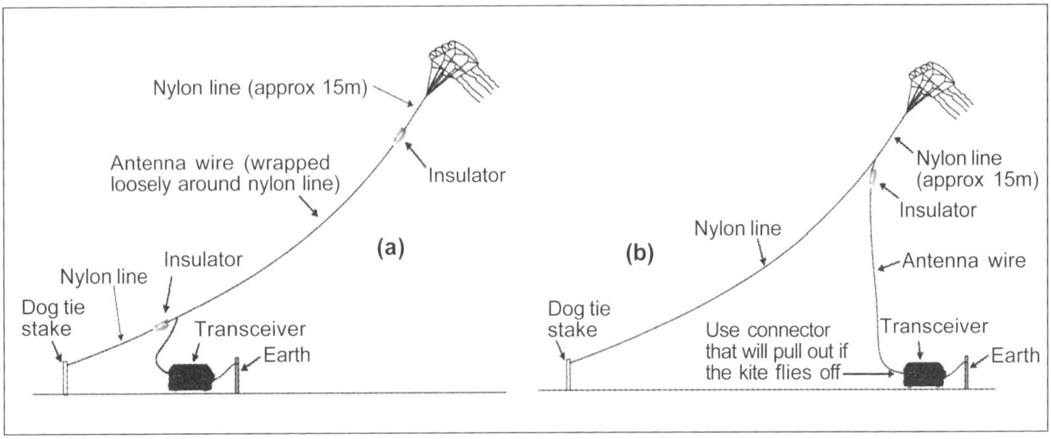

Fig 2.19: The two main ways of flying a kite antenna.

perfectly vertical antenna, if the wind drops a little you can end up with a lot of wire on the ground. **Fig 2.19(b)** shows this arrangement. Both arrangements work best when fed against radials laid out on the ground.

The bands where kites give the best results tend to be 80m and 160m. My own favourite antennas are a 5/8th-wavelength wire for 80m or a 3/8th-wavelength for 160m. If flown using the method shown in **Fig 2.19(a)**, they will (just) fit within the height restrictions.

Recovering a kite in strong wind can be difficult so to go prepared: use heavy gardening gloves for handling the kite line when it is in the air to avoid rope burns when recovering the kite in a strong wind. With a tethered kite, bringing it to the ground is sometimes easiest by running a carabiner along the kite line to gradually pull the kite down to the ground.

Don't fly your kite near power lines or where it will fly over roads. Remember that the wind can change direction! Be especially vigilant about lightning – check the weather forecast before you go out and cancel kite trips if thunderstorms are forecast.

Geoff Dobson, G8OFQ, used a Super Sled kite to raise a 240ft (73m) wire when he operated as HC8/G8OFQ from the Galapagos Islands, July to September 2015. Note that, unless specific permission has been obtained, the maximum height permitted in the UK is 60m.

Even if you take these precautions you can still run into problems with precipitation static: in bad cases the intensity of the noise caused by the precipitation static will drown out all but the strongest signals. There will also be large voltages present on your antenna that could damage your receiver (or you!) I have found that it is worst in summer conditions when there is very fine rain being blown by a brisk warm breeze.

The static can be discharged to earth by connecting a 1MΩ resistor between the end of the antenna and earth or by using an RF choke with an impedance of about 10 times the expected impedance at the end of the antenna. Neither will stop all the RF noise but they should protect you from serious damage. An earth system of one or more metal ground stakes and / or a few radials will be required to bleed the static to ground but it is safest to cease flying in such conditions.

The Civil Aviation Authority is the regulator for kites in the UK. No permission is required for heights up to 60m above ground level unless you are in an aerodrome traffic zone (ATZ), where the maximum height is 30m. The CAA can advise if your proposed location is within an ATZ. If you wish

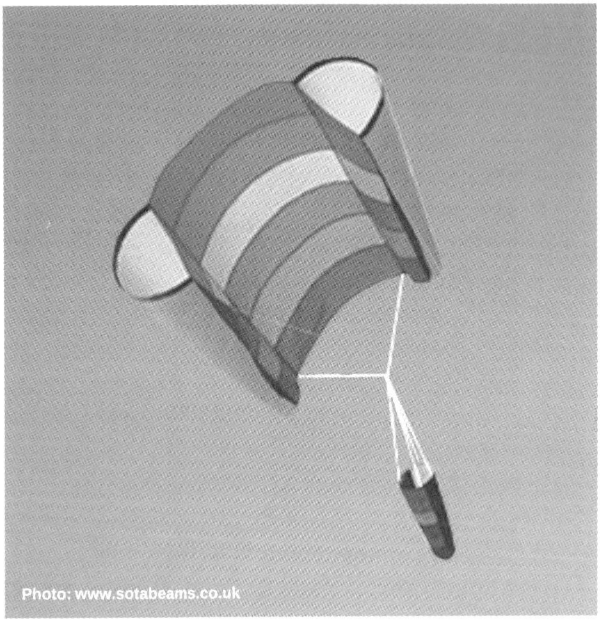

The Super Sled kite marketed by SOTABEAMS as the 'SOTAkite'.

to fly your kite higher, you must seek permission a minimum of 28 days in advance **[10]**.

Not mentioned in his 2008 article, Richard, G3CWI, now sells Super Sled kites such as the one used by HC8/G8OFQ through his SOTABEAMS company. Marketed as the 'SOTAkite', it has a Ripstock nylon flying surface and measures 168 x 100cm. Operating wind speeds are listed as 5 – 18MPH. The SOTAkite comes with a drogue tail, 200ft (61m) of 55kg Dacron braid on a reel, instructions and a kite bag. The SOTABEAMS website has a useful page of FAQs ***[11]*** *for those considering using a kit to support an antenna — Ed.*

REFERENCES

[1] 'The Miracle Whip – a Multiband QRP Antenna', Robert Victor, VA2ERY, *QST*, July 2001.

[2] Steve Yates, AA5TB, website (end-fed half-wave): www.aa5tb.com

[3] JAB Electrical Components: www.jabdog.com

[4] G0KYA website (end-fed half-wave vertical design): g0kya.blogspot.com/2010/11

[5] 'An Experimental Look at Ground Systems for HF Verticals', R Severns, N6LF, *QST*, March 2010.

[6] Islands On The Air (IOTA): www.iota-world.org

[7] *EZNEC* antenna modelling software by Roy Lewallen, W7EL: www.eznec.com

[8] Spiderbeam: spiderbeam.com (or, in North America, spiderbeam.us)

[9] Richard Newstead, G3CWI: www.sotabeams.co.uk/about-us

[10] Guidance on the regulations for kite flying at a height of more than 60 metres above ground level: www.caa.co.uk/General-aviation/Displays-events-and-activities/Kites

[11] SOTABEAMS kite FAQs: www.sotabeams.co.uk/kites-faq

Chapter 3

HF Dipoles and Doublets

PORTABLE 'ZIP CORD' ANTENNAS & FEED LINES

William A Parmley, KR8L
QST, March 2009

We start this chapter with what might be thought of as the simplest possible antenna – you don't even need any feeder cable! But there's a lot to consider, as William, KR8L, points out. He wrote, "Because of my interest in portable low power operation on the HF bands (often referred to as 'HFpacking'), I decided to look into the possibility of constructing lightweight, portable antennas and feed lines using commonly available 'zip cord'.

William acknowledged that the usefulness of 'zip cord' as a feed line for the higher HF bands has often been questioned, but stated that "Although my results are in general agreement with the earlier tests, I believe I can show that with the proper selection of material and careful deployment, feed line losses can be minimised on the higher HF bands, making zip cord dipole and feed line combinations satisfactory performers." This is especially true for portable, low-power operations in which some compromise may be acceptable in the interest of saving weight and bulk. Other potential uses for this type of antenna and feed line combination might include field day

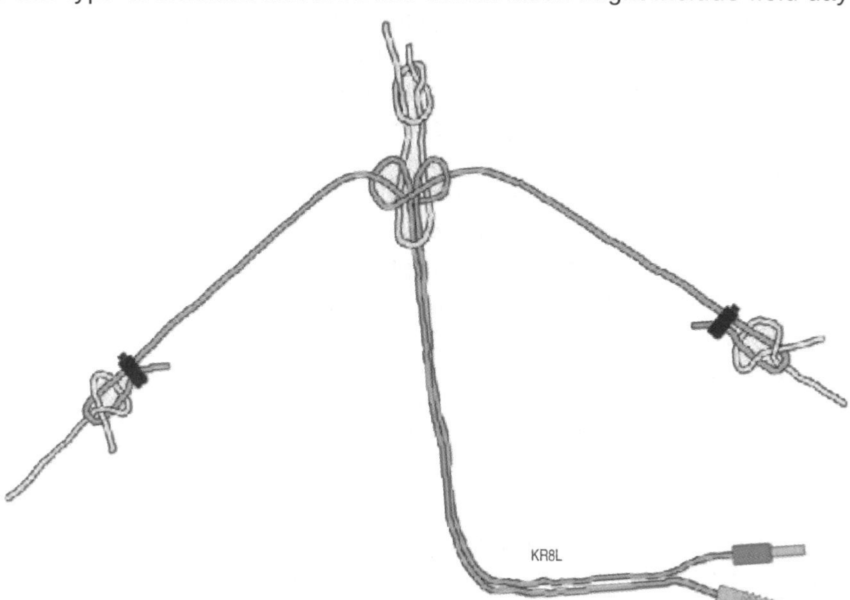

activities or 'stealth' situations in which the operator needed an easily-stowed antenna that could be erected quickly they wanted to operate.

He continued: My plan was to use zip cord as a half-wavelength feed line so that the dipole's impedance would be repeated at the transmitter end of the feed line. The impedance terminating a feed line is repeated every half wavelength. Because of this, the characteristic impedance of the feed line would be of secondary importance. Also, because the resulting feed line would be relatively short, losses would be minimised. This meant that I would not be able to erect the dipole at 'optimum' height, but since the goal of this project was to create light weight, compact, pocket sized antennas and feed lines that could be used in the field with temporary supports, antenna height was not a primary consideration.

After testing a couple of different samples I settled on RadioShack No. 278-1385, #22 speaker wire as a likely candidate. This wire is sold in 100 foot rolls and has a clear insulating jacket.

Measurements were made on an Autek Research Model VA1 Vector RX Antenna Analyst. Characteristic impedance was a property that could not be measured directly with the VA1, although some calculations and educated guesses could be made.

To begin I measured the centre-to-centre distance (S) between the conductors of RadioShack No. 278-1385 speaker wire as 0.082 inches, and found the conductor diameter (d) of #22 wire listed in *The ARRL Electronics Data Book* as 0.0253 inch. For parallel conductors with air dielectric the characteristic impedance is given by:

$$Z = 276 \times \log (2S/d) \text{ or:}$$
$$Z = 276 \times \log (2 \times 0.082 / 0.0253) = 224\Omega$$

...for an air dielectric. The plastic insulation on the wire should reduce the characteristic impedance by some amount, and although I didn't have a way to measure this effect I did come up with a way to estimate it. Here's where the 'educated guess' part comes in. First I did a similar calculation for nominal 300Ω twin lead, yielding a result of about 400Ω, meaning that the polyethylene insulation must reduce the calculated value by a factor of about 0.75. *The ARRL Antenna Book* says that for an insulated line, the characteristic impedance can be calculated by multiplying the 'in air' value by the inverse square root of the dielectric constant for the particular insulation being used. For polyethylene the dielectric constant is 2.3, so the adjustment factor should be about 0.65.

However, with typical 300Ω line the insulator is very thin so that the field between the conductors is only partially in the insulation and partially in air. It seemed reasonable that the adjustment factor (0.75 in this case) would fall somewhere between pure insulation (0.65) and pure air (1.0).

The type of insulation used on the RadioShack speaker wire isn't specified, but I made the assumption that it is polyethylene or a similar material (this seemed safe since many plastics have a similar dielectric constant of about 2.3). In addition, since the insulation between the conductors of the speaker wire is thicker (perpendicular to the plane containing the two wires) than the insulation for 300Ω twin lead, it seemed reasonable that it must have a greater influence on the characteristic impedance since more of the field between the conductors will pass through the insulation and less through the air. Based on all of this I assumed that the adjustment factor for the characteristic impedance would probably be closer to 0.65 than to 0.75, so the characteristic impedance for this line might be:

$Z = 224 \times 0.65 = 145.6$, or about 150Ω.

Again, this value is not particularly important, but keep it in mind and we will use it later to estimate line loss.

The velocity factor (VF) of the line was measured using the VA1. The technique involves shorting the end of the line, then sweeping the instrument over a range of frequencies to find the lowest impedance at several points. The lowest frequency will be the frequency at which the line is a half wavelength, the next will be two half wavelengths, etc. The velocity factor can then be calculated by the ratio of the physical line length to the value of a half wavelength in vacuum at the particular frequency (given by the formula $L = 492/f$, where L is in feet and f is in MHz). The results for my roll of No. 278-1385 speaker wire near four amateur bands are shown in **Table 3.1**.

Since I planned to build my antennas mostly for the 20m band and above, I chose to use 0.70 as the velocity factor of this line. (I could just as well have said that I picked 0.70 because it is a round number or because it is approximately the average of the four readings. I think these results are amazingly consistent considering the use of a consumer grade handheld instrument. There is only a 4% variation in measured velocity factor over a frequency range of about an order of magnitude.)

The attenuation or line loss was also measured using the VA1. For this calculation the series of minimum impedance measurements taken during

Frequency (MHz)	Velocity Factor (VF)
3.31	0.68
6.75	0.69
13.67	0.70
27.77	0.71

Table 3.1: Zip cord measured velocity factor.

Freq (MHz)	RS278-1385	RG-174	RG-58
3.31	0.97	2.7	0.8
6.75	1.48	3.3	1.2
13.67	2.39	4.0	1.6
27.77	3.41	5.3	2.4

Table 3.2: Calculated attenuation of zip cord compared to small coax, dB/100 feet.

the frequency sweep were applied to the following formula from the *VA1 Instruction Manual*:

$$\text{Loss} = 8.69 \times Z_{MIN}/Z_0$$

...where Z_0 is the characteristic impedance of the line.

The calculated loss is given in **Table 3.2**. Values for RG-174 and RG-58 (from *The ARRL Antenna Book*) are listed for comparison.

Now, let's think again about how I planned to use this feed line (a single half wavelength between transceiver and antenna). As the frequency increases so does the loss, but the length of a half wavelength of feed line decreases. As a result, the feed line loss remains less than 1dB as we go up in frequency. In fact, the loss of a half wavelength line decreases from about 1dB at 80m to about 0.5dB at 10m.

If the estimate of a 150Ω characteristic impedance is correct, then the SWR on the line will be about 3:1, which introduces at most an additional 0.7dB of loss (as read from the graph of mismatched line loss in *The ARRL Antenna Book*). Total feed line loss would then be about 1.5dB, and certainly less than 2dB. As you can see from the table, the line is slightly more lossy than RG-58. It is closer in size and flexibility to RG-174 and is much less susceptible to damage from bending and rough handling than is coaxial cable, an important consideration given the intended use. It might be reasonable to make the feed line a full wavelength long on the higher bands in order to increase the antenna height. Whether the increase in height would offset the increase in feed line loss probably depends on the individual installation.

Having characterised the wire as completely as possible I next wanted to check my calculations by building and testing some examples. First I calculated the antenna length using the formula $L_A = 234/f$ and then calculated the feed line length using the formula $L_F = (VF) \times 492/f$. I used the quarter wavelength formula for the dipole since the wire was going to be 'unzipped' to form the two halves of the antenna and applied the standard 5% reduction in length for 'end effects'. I used the half wavelength formula for the feed line multiplied by the measured velocity factor. A single piece of speaker wire of length $L = L_F + L_A + X$ was then cut, where X had a couple of feet added to the length for margin of error. The feed line length (L_F) was measured and marked, and the remainder of the wire, including the extra length, was unzipped to make the dipole section. An electricians' knot (see **Fig 3.1**) was tied at the junction of the dipole and the feed line.

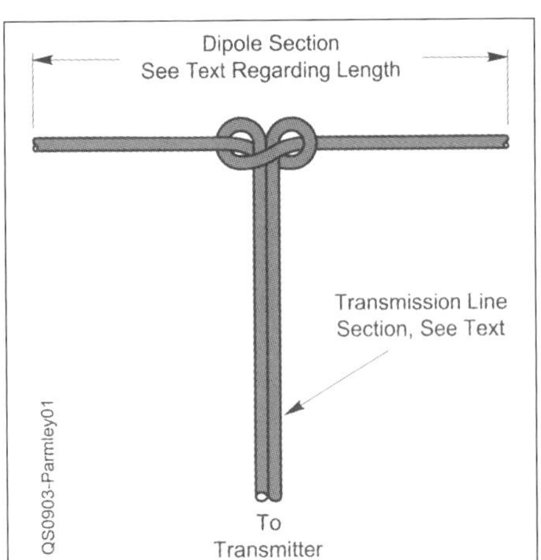

Fig 3.1: The centre of the antenna section can be secured with an electrician's knot as shown.

At the transmitter end of the feed line I unzipped the wire a couple of inches and attached a banana plug to one side and an alligator clip to the other. The banana plug fits perfectly in the centre conductor of a transceiver's SO-239 coax connector, while the alligator ('crock') clip makes a convenient way to attach to the transceiver's ground connection, as shown in the photo. I completely ignored the issue of feeding a balanced load with an unbalanced source here, but if you think about it, this is not an uncommon practice for simple wire antennas, and hams have been doing this kind of thing for ages.

Rear of transceiver showing banana plug and crock clip lead connections.

We're just more accustomed to seeing this where the feed line (coax) meets the antenna (dipole) than where the transceiver meets the feed line. For low power applications I have not found this arrangement to be a problem. For those who are concerned about this, a few turns of the transceiver end of the feed line can be wound through an appropriate toroidal core to make a 1:1 transmission line transformer. I did test the transformer configuration briefly and found there was no difference in performance, feed point impedance or SWR.

For the initial setup of the dipole I measured and marked the calculated antenna length on each leg and folded the wire back on itself at this point and taped it in place. Later I found that a spring compression cord stop of the type found on drawstrings on jackets and other items of clothing worked much better than tape, especially when the length of the dipole legs needed to be adjusted. These should be available in most fabric stores. In addition to shortening the dipole while leaving the option of lengthening it without subsequent splicing and soldering, I found that this technique created a convenient attachment loop at the end of the wire. Finally, I tied a piece of light nylon line to the loop on each end of the dipole (no other insulator was used) and proceeded with deployment and testing.

After building antennas and feed lines for 30, 20 and 17m, the initial testing was done by installing the antennas in an inverted-V configuration with the apex at about 20 feet. This was done using either a telescoping fishing pole, or by tossing a line over a tree branch and pulling the dipole up with that. The ends of the dipole were brought down to 6 to 8 feet off the ground and tied off with nylon line that was then tied to tent stakes. The dipole was pruned to resonance using the VA1 by changing the fold point at the end. The extra wire was left in place and was not trimmed off.

Complete battery-powered QRP station using a zip cord antenna and feed line.

In practice I found that once the antenna was tuned to resonance it was possible to adjust and optimise the feed point impedance by changing both the horizontal and vertical angles between the two legs. In my particular outdoor installation the best match was found with the dipole legs arranged at a horizontal (azimuthal) angle of between 90° and 120°. As should always be done with parallel wire feeders, I made an effort to keep the feed line clear of other objects and equidistant from both legs of the dipole to the maximum extent practicable.

I have used these antennas and feed lines for SSB, CW and PSK31 at QRP power levels in indoor, backyard and backpack portable situations. The photo above shows a self-contained portable station. Portability is excellent, deployment is simple and on-air performance seems to be very good.

Note that some zip cord may be significantly more lossy than the type that I used, so it is important to make measurements before committing to a particular type.

PORTABLE 5MHZ NVIS ANTENNA

Chris Saunders, G4ZCS

RadCom, March 2012

Sometimes HF Near Vertical Incidence Skywave (NVIS) propagation is required in order to make local contacts from locations where VHF or UHF working is impossible or impractical. Chris, G4ZCS, designed a 5MHz NVIS antenna for use on portable emergency communications exercises at a time when the 5MHz frequencies were only available to UK Full licensees by means of a 'Notice of Variation' (NoV) to their licence.

Chris wrote that "In early summer 2010 Ken, G3WYN, started experimenting with a very low wire antenna for 5MHz. Having seen this unusual antenna under test in his garden I decided to have a go myself, as I had been looking into producing an easily repeatable antenna for possible emergency communications use."

G3WYN's original design consists of two 9.2m (38ft 2in) lengths of PVC insulated 7/0.4mm wire with a dog bone insulator at the feed point, connected to the coax feeder via a 1:1 air cored balun attached to a low ground stake. The wires were extended as a 90° 'V', pointing approximately south-east, supported on upside-down plastic flower pots that were spaced out on the lawn grass.

At the far end, the wires were anchored and kept taut via another dog bone insulator and ground stake. The wire lengths were cut for resonance, in the mid band slot that our NoVs allow. An antenna analyser was used, obtaining an indication of 50Ω on the display at resonance.

Signal reports indicated an effective antenna, with its signals from Sussex regularly generating reports from as far away as Scotland.

Our local emergency services proposed a major railway evacuation exercise deep in the valleys of the South Downs, to which our club was invited to test communication links. This prompted me to experiment with an easy to erect emergency antenna, described here. My version broadly follows Ken's design but supports the wires using animal electric fence supports, made largely of plastic but with a metal spike at the base. The theoretical height of this antenna should be about a quarter of a wavelength above ground (14 metres), however, a practical antenna needs to be a bit lower. A height of just a metre or so is ideal to make the antenna truly portable in a small car, easy to erect single handed in all weathers – and in the dark.

The photo below shows the complete radio station kit, including my Elecraft K3, a 24Ah gel battery and a small operating table.

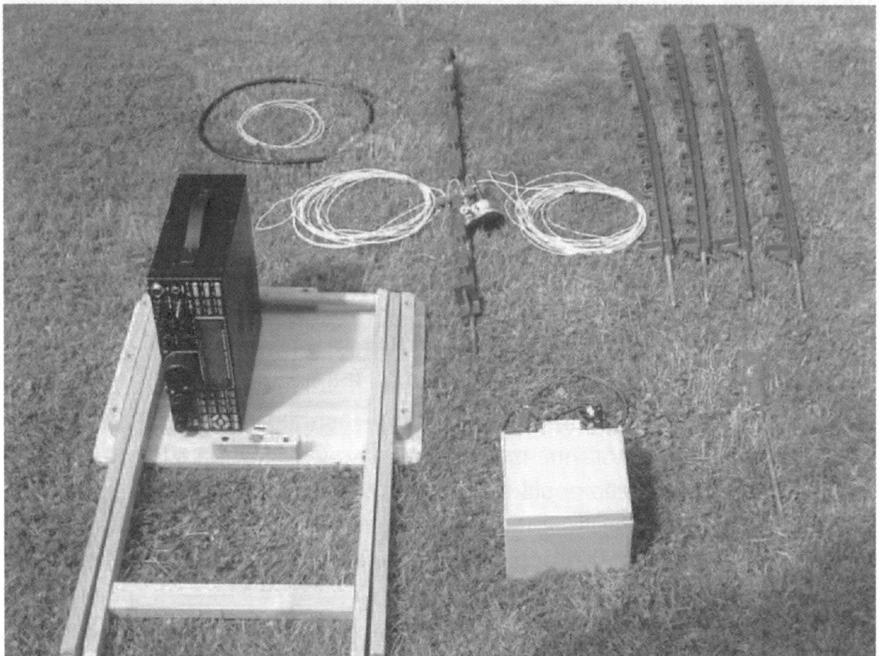

The basic portable 5MHz NVIS antenna set-up. Clockwise from top left: coax and earth cable, support post with coiled antenna elements and feed, four additional support posts, long screwdriver to act as station earth point, 12V gel cell, folding table and Elecraft K3 transceiver.

General view of the antenna and portable operating position. The set up time is less than five minutes.

The photo above shows the operating arrangement.

The 'T' piece at the centre of the dipole is a piece of engineering plastic from the (Bluebell) railway track scrap bin. I tested it in the microwave for a few seconds, alongside a cup of water – the plastic didn't get warm, so I decided it was OK for RF.

The balun is mounted in a 1oz tobacco tin. It is a dual ratio 1:1 – 4:1 transformer type toroid wound on a 40mm (1½in) diameter ferrite ring. There are 14 trifilar turns of enamelled copper wire, separated from the ferrite by a layer of PTFE tape. The unbalanced terminations are bought out of the enclosure using phono sockets – a phono to SO-239 adapter is used to couple the feeder.

Using 'normal' free space calculations, the length of a 5MHz half-wave dipole antenna should be in the region of 28m. However, the proximity to the ground caused resonance to occur at a much shortened element length. My elements are each 9.13m from the feed point insulator, plus 10cm tails to the balun. The tails are 10cm apart on the insulator, making the total wire elements 9.23m; the antenna is 18.36 metres overall when laid out straight.

Using the various clip positions on the uprights I tested the antenna from 23cm (9in) up to 84cm (2ft 9in) above ground. The optimum seems to be at around 23 – 30cm (9 to 12in), giving a resonance centred at 5.35MHz and a resistance of just over 35Ω.

Initially, the radio picked up a lot of noise, but by connecting an earthing spike (the screwdriver in the photo opposite) to the rig and keeping the clay ground wet, the noise level was reduced considerably.

I think this antenna makes an ideal low-band system for quick and repeatable operation by a single operator for emergencies and field day operations. I decided to go out portable for further tests, firstly at Jack and

Feed arrangement. The tobacco tin contains a 1:1 / 4:1 balun.

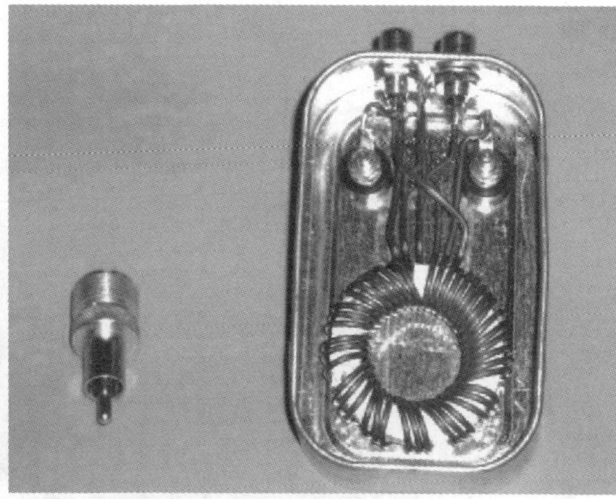

Detailed view of the balun (see text).

Jill Windmills, high up on the South Downs. Here, the ground conditions were summer dry chalk, 210m (700ft) ASL. The antenna was run out in a straight line about 30cm above the ground, running roughly north-south. The whole station took less than five minutes to unpack from the car, assemble and tune up. The antenna was quieter up on the Downs and produced much better results than tests from home. With the evening sun still up (late June 2010), initial contacts were made with Ken, G3WYN. He was still using his flowerpot antenna and the 7 mile line of sight contact was via ground wave.

I then contacted several stations in East Anglia, all coming in at about 58 and 59. Later, more locals appeared from within a 20-mile radius, suggesting either good E-layer propagation or NVIS via the F2 layer.

After the sun had set more contacts were made, this time with the far south west, from mid-Devon down to Cornwall and, in the other direction, into Essex. Most of these later contacts produced reports of S9+20dB, with me using transmit powers in the range of 25 to 50W.

The following June, I operated again from the same site, so the ground conditions were the same. This time, to validate the performance results of the antenna, it was orientated east-west, to see if there were any appreciable differences. This way, the broadside signal was launched off the north face of the hill. Unfortunately it was some time before we realised that the antenna was only about one metre from a metal mesh sheep fence, buried in a hedge, which was possibly acting as a reflector.

The first contact was again with Ken, G3WYN, this time with his flowerpot antenna in a straight line. He was as strong as expected for the 7 miles line of sight. This contact was followed by G4PTJ, at just 8 miles but hidden behind an iron stone ridge. We had 57 reports each way. Next came G0WGP at 3.5 miles, line of sight, at 59+.

At this point some longer distances were worked. Colin, G0DMB, gave me 58 to 59 in Leicester; Jim, G4REK, 59 in South London; Tony, G4BFS, heard me 58 in Norwich; Paul, G0HNW, 59+ in Derbyshire; Dave, G3VSJ,

General view of G4ZCS/P operating in the quarry. The antenna is close to the trees; the top of the cliff can be seen through their upper branches.

55 in Norfolk; Ron GW6ZDH, 57 to 58 in Conwy, North Wales and, finally, Paul, M0YMJ, gave me 59 in Canterbury.

Just as it was getting very dark a CQ call from OZ7AE was heard, however, our time had run out and we had to pack up the station, so there was no time to work Benny (who was about 56).

Probably the most important test of all was in October 2011. The test was to operate from a deep valley type site where there is no line of sight path available. Finding one where there might also be mobile phone coverage (for talkback) was a major problem. However, an abandoned chalk quarry nestling deep in the South Downs proved to be the ideal spot. It is horseshoe shaped, well over 30m (100ft) deep (and as a bonus is home to an excellent pub!)

The station was set up as before, with stakes in the soft turf, but they would not go in very deep because of the solid chalk beneath. There were tree canopies above, and the cliff face was about 50ft away. **Photo 6 and Photo 7** give a good idea of the operating conditions. The opening of the horseshoe shaped bowl was 180° from the RF path.

Arrangements had been made for G3WYN (at 12 miles), G0WGP (7 miles) and G7TMR (9.5 miles, as a listener) to be on channel FL at 1330z. On that day there was a SFI of 147, the A index was 4 and the K index 2. On turning on the radio, some signals could be heard; after a quick tune up, G3WYN was heard at 58. A call was made, and he gave me a 55 reply from my 50W initial power setting. G0WGP reported 53 with high background noise. His report for a 25W transmission was 56. The quarry site attenuated the background noise affecting the other two stations to a level of lower than S2. A relayed report was received from G7TMR of 57.

A number of CQ calls were made. However, there were no other stations active at the time of the tests.

NVIS was proved to be the transmission mode; no VHF path was available even to the local repeater network and only a weak mobile phone signal was obtainable at the quarry entrance (presumably, from the nearby town of Lewes).

A low strung antenna wire is an obvious trip hazard; small plastic bunting flags were used on the wire in public areas. (Raising the height of the antenna would have been safer in public areas; however, this would have caused complications in erecting and stabilising the elements in adverse conditions). Alternatively, a high visibility tape could be strung along the top of the supports. There is also a risk of RF burns from accidental contact with a transmitting antenna; in our tests, a good visual lookout was observed and transmissions were stopped if anyone looked like approaching the antenna.

If you choose to have a vehicle near the antenna it is well worth checking the manufacturer's specification for the maximum RF power to be used near it. When I transmitted near the car I limited my power to 50W, which is less than the limit stated in the car's handbook.

Although working with limited data, my initial conclusions are that this antenna is successful for 5MHz contacts from 3 to 250 miles. The line of sight performance is best off the side of the antenna, whilst the NVIS is good at 10 to 80 miles. Lower angle lobes to the side of the antenna are good for ranges 100 to 200 miles from an elevated test site.

Its theoretical efficiency must be in doubt due to its proximity to the ground; however, the results speak for themselves – it works! On the elevated dry chalk, the nominal ground plane could be considered to be a long way down, making the antenna behave more like a free space antenna. To prove this, another test will be needed after a period of prolonged rain. It might even help to get the support spikes to penetrate the ground!

It has proved to be ideal for a quick set up and take down, possibly under emergency conditions and even in the dark. It is easy to transport and seems to have repeatable performance. The best bit is that mine cost me less than £10 (for the five fence stakes).

It is worth remembering that if the antenna is to be used in an urban environment some alternative to the ground spike will be needed, as these agricultural fence posts are not designed to penetrate concrete!

It might be possible to scale the design up or down, for 40 or 80m, to test its performance on these bands. Ken has used his successfully on 40m by folding back each leg on itself to achieve the higher resonance.

This antenna, on this band, can be recommended for emergency communications from almost any rural site over a typical county-wide distance. It is also plain to see why the armed services are so keen to use frequencies in this part of the spectrum!

THE 40m 'SKY-PAINTER'

Ian Thomson, GM0URD
RadCom, July 2014

The Sky-Painter was the name given to this quick-to-assemble rotatable dipole antenna designed by Ian, GM0URD. He wrote that he wanted to construct a simple yet practical dipole antenna for 40m for use in field day events that would would fit into the back of his small car.

The Sky-Painter is made from two Watson WHF-40 single-band mobile antennas mounted horizontally with a home-made support and fed with a balun. The name, 'Sky-Painter', came about because the mast section consists of a paint roller extension pole, fitted on the top of a window cleaner's fibreglass telescopic extending pole.

The two mobile antennas are mounted back-to-back on separate mounts, forming a dipole. The mounts are bolted to each other so that they clamp on to the window cleaner's pole that forms the mast section, as shown in

General view of the fully erected Sky-Painter portable 40m dipole antenna.

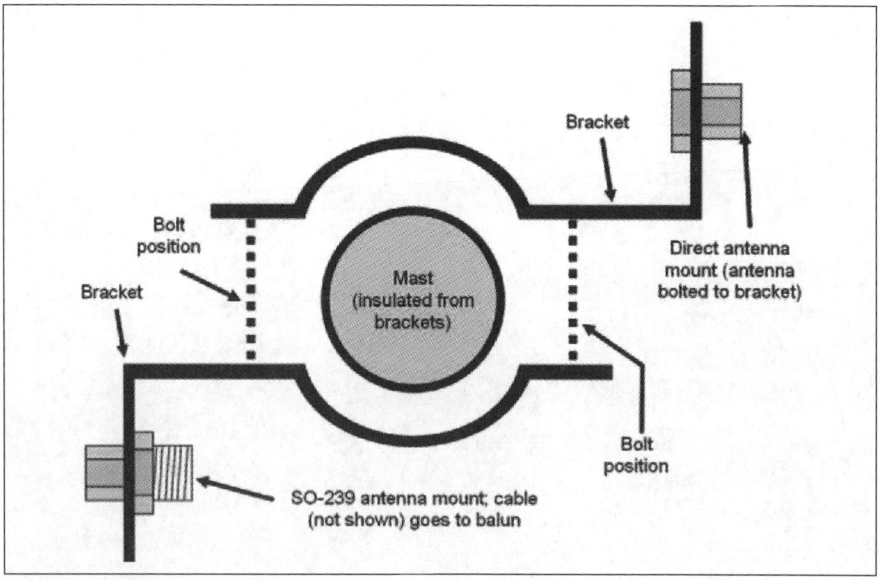

Fig 3.2: Two brackets bolted either side of the mast provide mountings for the elements.

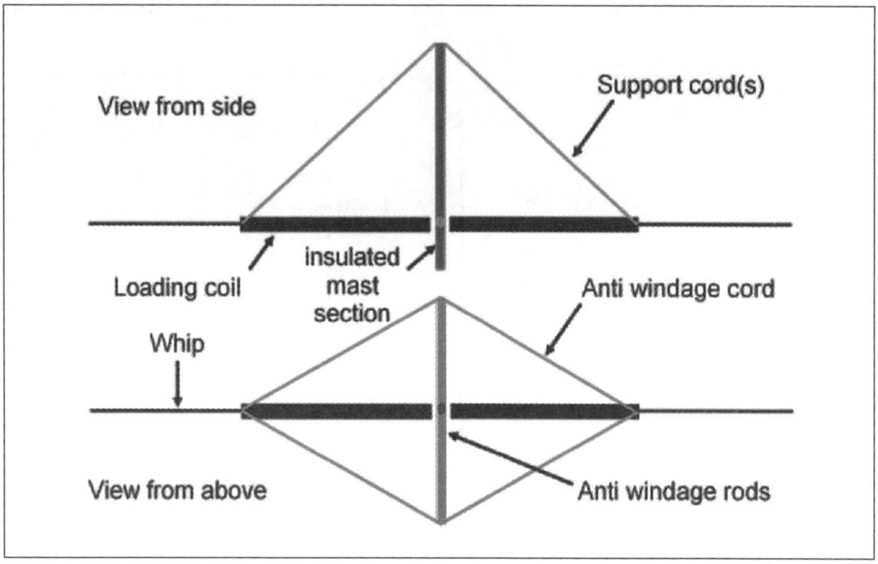

Fig 3.3: Basic principle of the antenna guy system. Upper image: resisting gravity. Lower image: anti-windage system.

Fig 3.2. In order to keep the mobile whips horizontal without putting excessive strain on the mounts, a system of guys is employed, dropping down from the top of the mast. Windage would also cause stress, so an additional horizontal guy system is employed, as shown in **Fig 3.3**.

In order to reduce current flow on the outer of the coax, a balun is included immediately before the feed point. This consists of five turns of coax looped

Detail of the balun.

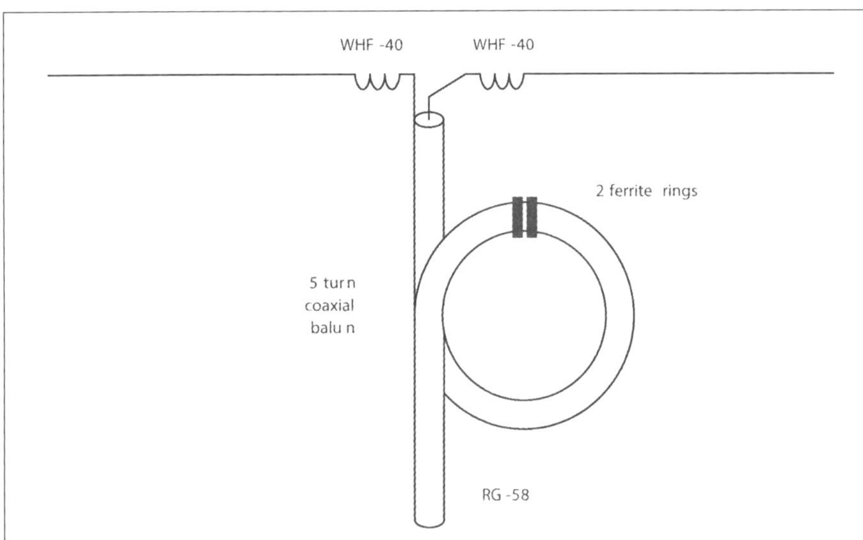

Fig 3.4: Electrical arrangement of the antenna and balun.

through two glued-together ferrite rings, the whole of which is mounted on a kitchen food-cutting board, which can be seen in the photo above. **Fig 3.4** shows the electrical arrangement of the balun.

The base section is constructed from a modified, folding work bench, to which the fibreglass mast section is attached. The whole Sky-Painter folded up ready for transport to the portable location is shown in the photograph on the following page.

General view of the complete antenna system packed ready for transport to the field day site.

The centre of this pole has a hollow hexagonal plastic section, which protects the coax. The coax goes up through the centre of the mast and out through a grommet in the centre of one of the antenna element support brackets. This bracket had to be modified slightly by drilling out and rounding off a hole in its centre for the coax to pass through from the mast.

The balun support is made from a kitchen food-cutting board, with holes added to reduce any windage effect. The home-made anti-windage support, Watson WHF-40 antenna sections and the upper antenna support bracket, which is constructed from a paint roller extension pole, can all be seen in the photo above.

In order to provide a reasonably sturdy base for the Sky-Painter, a typical folding DIY work stand was used. Alternatively, a galvanised car stand-on type support bracket could be used, with a short piece of scaffold pole attached for a more portable base.

The choke balun converts the 50Ω unbalanced feed to balanced 50Ω for the antennas and is constructed from five turns of RG58 feeder, passing through two 25mm OD ferrite rings that are glued together. The coiled coax has a centre coil diameter of 140mm. This is mounted on a kitchen cutting board by using nylon cord or cable ties. The back of the balun support board has plastic spring type pipe clips at the top and bottom for clipping it to the mast, to prevent it from blowing about in the wind. Vertical support is provided by loops of nylon cord that go between two plastic kitchen shelf support blocks on the cutting board to two small metal shelf brackets attached to the antenna mounting bracket. Grommets are fitted to the metal brackets to prevent damage to the cord. The dual ferrite rings have three medium cable ties around them to help protect them from any possible chipping while on the move.

The mast was a 'Harris Poz Lock' fibreglass window cleaner's extension pole. These are available in two sizes, both of which extend to about double

Further detail of the balun mounting.

their collapsed height. In practice, the larger pole was too big to fit inside the writer's small car, so the shorter extension pole was used. The Poz Lock feature of the pole locks into spaced holes that extend up the length of the top extending section, stopping it from rotating about.

The pole is secured to the base using clamps constructed from 54mm car exhaust clamps with the back of the clamps replaced by strips of rubber, as shown in the photo below.

The Watson dipole elements required extra hook type supports to be fitted so that the support cords could be attached. Originally the nylon string was going to support the ends of the steel whips but, in practice, they bowed back on themselves and it was found that the steel whips themselves actually required no support at all. The support is now connected to the middle section, as can be seen in the lead photo.

Before connecting to the transceiver it is sensible to first check that the antenna elements are roughly in the centre part of the band where you wish to operate. GM0URD used a length of about 70ft of RG-58C/U 50Ω feeder and estimated the loss to be less than 1dB at 7MHz, as this coax has a loss of about 1.4dB per 100ft at 10MHz. He wrote, "I was not too fussy about this, as after all it was a field day antenna". The initial tuning was done with the mast lowered. Remember to balance the lengths of the stainless steel whips! For the first test, a dip meter indicated resonance at about 7100kHz, which was a great starting point. For final tuning the mast was raised to full height and the coax connected to the transceiver with an SWR meter in line but no ATU. Testing was done at reduced power to start with. At 7130kHz the VSWR was 1.1:1 in dry conditions but dropped to 1.3:1 in the wet, probably due to increased capacitive coupling between the elements and ground. The 2:1 VSWR

Clamping the mast to the support.

bandwidth in dry conditions was found to be about 84kHz. This is slightly better than the manufacturer's quoted typical bandwidth of 80kHz. When tuning the antennas it is wise to heed the manufacturer's warning about whip lengths: "If required, cut off the bottom of the whip to ensure that it does not extend into the dense part of the coil winding. Failure to observe this precaution will cause extreme heat damage and void any warranty."

The Watson WHF-40 antennas are rated at a maximum power of 200W PEP. The generally-similar ProAm PHF-40B 40m antennas could be used in place of the Watson WHF-40 antennas, but according to the manufacturer's figures are likely to give a bandwidth of nearer 60kHz, though with a power rating of 250W PEP. Do not mix antenna types, because mechanical balance is as important as electrical balance.

THE 'TEE PEE V'

Robert Giuliano, KB8RCO
QST, June 2008

The Tee Pee V is a simple dual-band antenna that can be set up by a single person. It was designed for use on field days and special event stations after Robert's local radio group found themselves at a location no available

The Tee Pee V is a simple dual-band antenna for Field Day operations named after its similarity in appearance to the Native American tent.

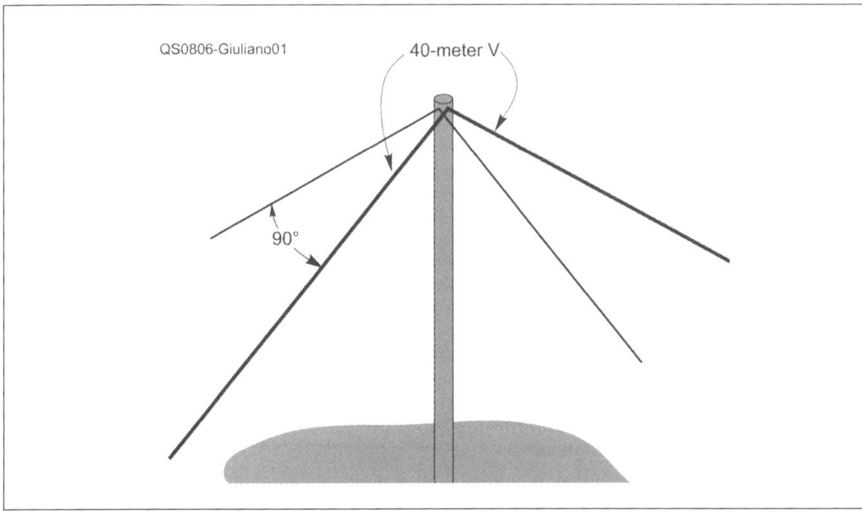

Fig 3.5: General view of the Tee Pee V.

supports for antennas. They designed a simple, self-supporting and effective antenna that could be put up almost anywhere and be easily transported.

They first tried a 20m inverted-V mounted on a 22ft (6.70m) fibreglass pole. The Tee Pee V concept came from Robert, N8RGF, as he looked up at an inverted-V and asked, "Why do we need four guys *and* the dipole wires?" Can they combine the concepts and get two band operation from one pole using the antenna as the guys?

Inverted-V dipoles for 20m and 40m were tried, as shown in **Fig 3.5** and **Fig 3.6**. Because the two antennas are perpendicular to each other, the coupling that often causes problems with multiband parallel wire dipoles is minimised, and *EZNEC* predicts that operation on each band is much as a separate inverted-V. No balun was used at the feed point but if you experience

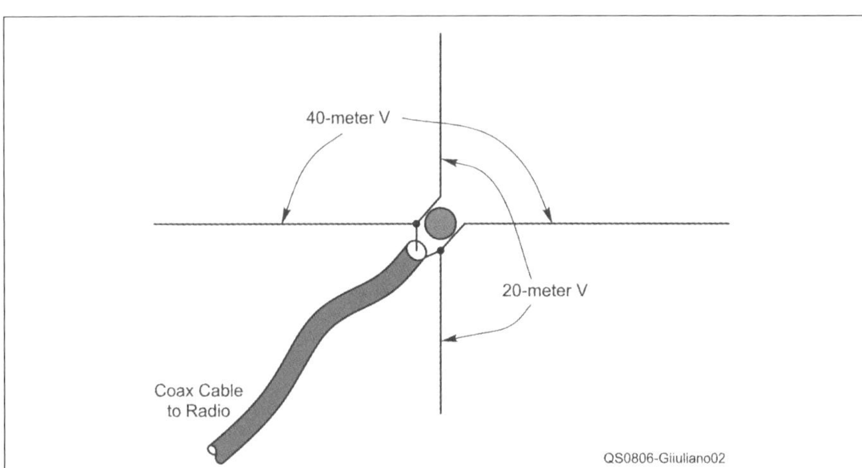

Fig 3.6: Details of the feed point connection arrangements.

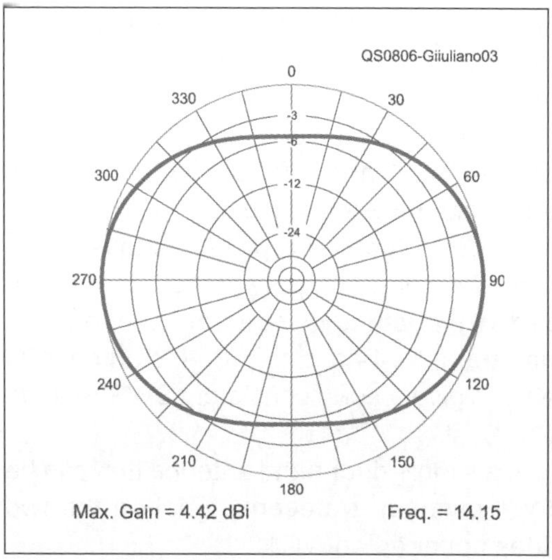

Fig 3.7: Azimuth plot of 20m Tee Pee V at elevation angle of 35°.

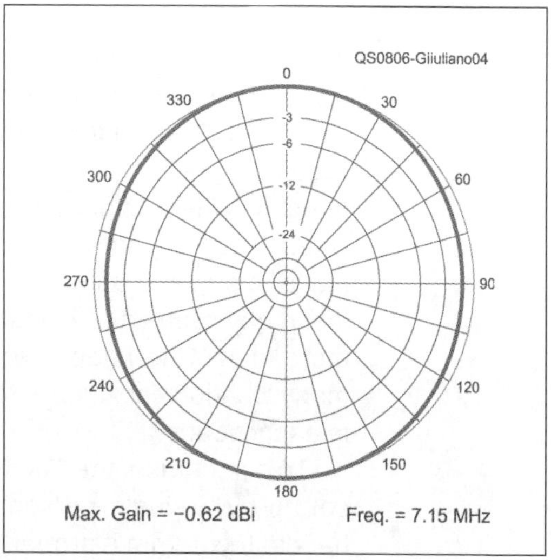

Fig 3.8: Azimuth plot of 40m Tee Pee V at elevation angle of 35°.

the effects of common mode currents on the outside of the coax you may want to add one. A choke as simple as half a dozen turns of coax, about 8in (20cm) in diameter, might be all you need.

The predicted performance over typical ground is shown in the azimuth patterns for both bands in **Figs 3.7** and **3.8**. As can be seen, the patterns are just what would be expected of any relatively low inverted-V, largely omnidirectional.

The antenna model determined the element lengths to be approximately 17ft (5.18m) for 20m and 33ft (10.06) for 40m. The total wire required would thus be about 100ft (30.48m). A little geometry was used to keep the angle with respect to the pole at 56.75°. This allows each guy stake to be the same distance from the pole and makes set up easy.

The distance the stakes need to be from the pole was calculated by multiplying the height by the tangent of the included angle (56.75°) and that came out to be about 33.5ft (10.21m). The 56.75° angle of each wire to the pole results in the angle between the antenna wires of 113.5°, nicely larger than the design goal for inverted-Vs of being greater than 90°.

The inverted-Vs were made from cheap 16 gauge speaker wire of stranded copper, one side tinned. A little extra wire was added to allow for trimming with an antenna analyser. After trimming the antenna measured a flat SWR across both bands. Robert, KB8RCO, wrote, "Next I connected my Yaesu FT-100D transceiver to the antenna and gave it a try. I check into a local net on 40m and received a good signal report (59) and I confirmed 40 was working. Now for the 20m side. I usually ask the Maritime Mobile Net operator, but the Coast Guard net had just started. I made a quick swing through the band to find a contact and got a 59 report from Georgia. This made me feel the antenna was usable."

A Mosley trap vertical was also set up and the Tee Pee V checked against that. Swapping the antennas back and forth, the signals were pretty close. But is the Tee Pee V repeatable?

KB8RCO wrote that he tried the same antenna at a different location: "I picked a spot for the pole and stepped off 11 paces (about 33 feet) in each direction where I pounded in a tent stake for each of the support ropes and wires. I tied slip knots around the stakes (adding rope slide stops would make this even easier), and raised the pole. The pole stayed up with the loose guy ropes and I was able to step out to adjust the ropes for a touch more tension to keep everything straight... I have set up this antenna in many locations now, and it appears to give a good consistent SWR reading in each location."

To summarise, the Tee Pee V is a simple dual-band antenna that can be set up by a single person and will give you a decent signal on the two bands. If you trim it properly it does not need an ATU.

Many years ago, when living in Scandinavia, I made a similar antenna for the 80m and 40m bands which was hoisted to the top of a wooden flag pole (without a flag). Inexplicably this antenna also presented a low SWR – and got out well too – on 20m. This is not guaranteed, though! — Ed.

DIPOLE CENTRE PIECE AND BALUN FOR LOW POWER PORTABLE OPERATIONS

Ron Davis, WB5TGF
QST, November 2013

Here's a clever way of using inexpensive PVC fittings to make a dipole centre piece which also includes a balun suitable for low power portable operating. While the references are all for American-made products and stores, the ingenious constructor in the UK or Europe should easily be able to source similar products.

Ron, WB5TGF, wrote: "I constructed a portable dipole antenna and its balun assembly after being inspired by a T-shaped PVC shower fitting that caught my eye in the hardware store. My design is for use in low power operation. I made my antenna for use on 40m, to complement a Yaesu FT-817 low power radio in my emergency communications 'grab-and-go' box, but you can build yours for use on any of your favourite bands..."

Needed for the assembly of my antenna was a 1:1 balun, a PVC shower Fitting and plug, wire for the dipole legs, and a coax receptacle. I used a panel type BNC receptacle, www.mouser.com part number 565-2451A, but you can use other coax receptacles as well. I wound the balun on a 0.5 inch diameter by 0.188 inch thick toroid from www.kitsandparts.com, part number T50-43, using #22 AWG enamelled magnet wire (www.radioshack.com, catalog number 278-1345). The toroid must be small enough to fit in the 0.625 inch diameter non-threaded opening of the 1/2 inch diameter 90 degree

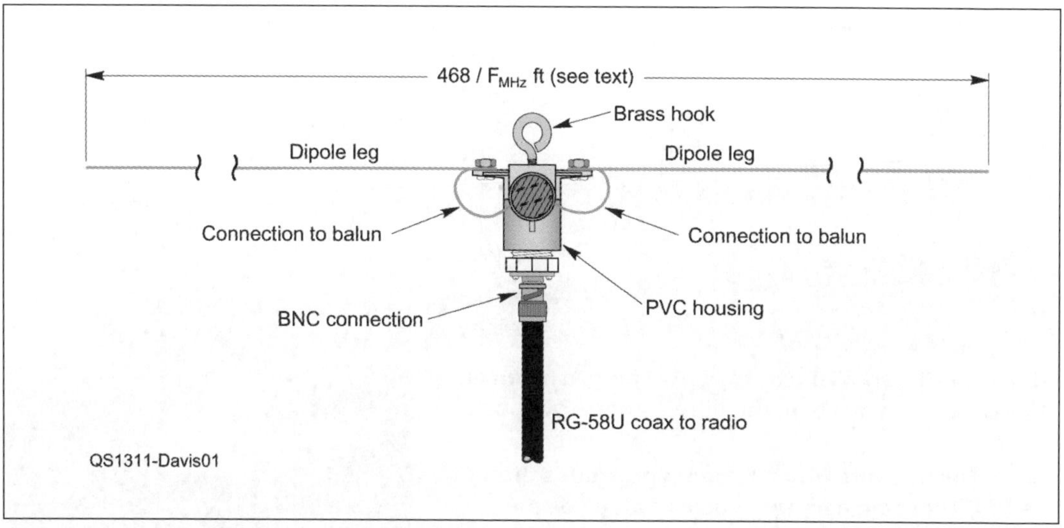

Fig 3.9: A PVC shower fixture component houses a balun and supports the centre of a dipole.

elbow CPVC shower fitting (from www.lowes.com, item 9734 model 53055N). You will also need a 1/2 inch PVC schedule 40 plug (www.lowes.com, item 23526, model 450005RMC).

The 1:1 common mode balun transitions from the unbalanced coax line side connection to the balanced antenna dipole side connection. I folded a 24 inch (61cm) length of #22 AWG enamelled wire in half and fed the folded end of the wire through the centre hole of the toroid, leaving leads a few inches long. I then pulled the folded end around and pushed it through the hole in the same direction, repeatedly, until I had wound nine turns of bifilar wire. I made the last turn come out on the same side as the starting two wire leads. The starting two wire leads are the balanced side of the balun, and the folded end of the bifilar wire is the unbalanced side of the balun. Remember that this balun and fixture were designed for low power operation.

The PVC shower fitting has a threaded and a non-threaded opening. I drilled a 1/16 inch hole in each side near the bottom of the non-threaded opening to pass one each of the two wires from the balanced side of the balun. I drilled a 1/4 to 3/8 inch hole centred axially in the PVC plug fitting.

I centred the BNC receptacle on the end of the PVC plug, and using the BNC receptacle flange holes for a pattern, drilled holes which I tapped for screws to hold the BNC receptacle to the PVC plug. I made a slot for one wire on the unbalanced side of the balun using a hot soldering iron tip. I used needle nose pliers to push each of the balanced side wires out the holes I drilled in the unthreaded part of the fitting. I pushed the folded end wires out the threaded part of the fitting, pushed the balun down into the hole of the PVC shower fitting while pulling gently on the two balanced side wires. The balun should fit snugly in place. I passed the folded end of the wire through the PVC plug fitting and screwed the PVC plug into the PVC

Above: A PVC shower fixture, a PVC plug, a balun and BNC receptacle make up the dipole centre support.

Right: The one wire from the balun protrudes through the BNC receptacle flange. You can also see the balanced side wires at the PVC flange.

shower fitting. (It does not have to be very tight. You will use silicone sealant to waterproof everything.) See the photo above left.

Next, I installed the BNC receptacle. I cut the folded end of the wire and pulled one wire over to the side. This end will connect to the BNC receptacle flange. I cut the other end off, leaving about 3/4 inches of wire protruding from the fitting. Be careful here and do not cut it too short. With a sharp knife I scraped the enamelled coating from the wire and tinned the scraped portion of the wire with my soldering iron. Next I soldered the tinned wire to the centre conductor of the BNC receptacle.

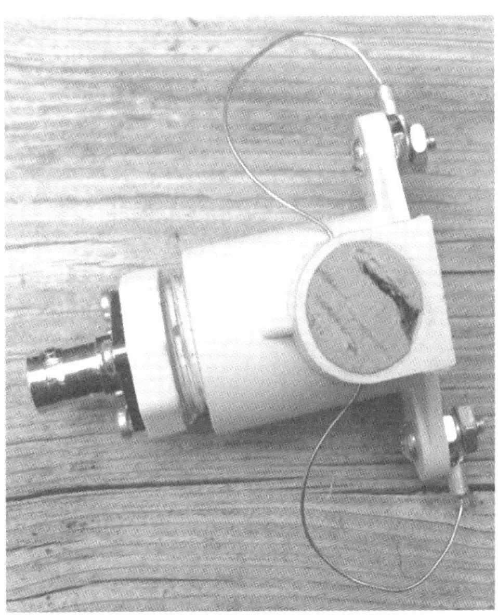

The assembled centre section for the dipole.

You can attach an optional hook on top of the PVC fixture.

I fed the remaining wire up through one of the BNC receptacle flange holes and gently pushed the receptacle down until I could get a screw started, but I did not tighten at this point. I put two more screws into the BNC receptacle flange, leaving the hole with the wire sticking out until last, as shown in the photo above right. I scraped the coating from the end of the ground wire where it would go under the screw, and wrapped the wire around the last screw after starting it. Finally, I tightened all four flange screws.

I installed two small bolts with washers and nuts in the PVC shower fitting flanges and, after removing the enamel coating from the two balanced side wire ends, I secured one wire under the nut on each side of the PVC flange. A second nut secures the dipole antenna wire. You can solder terminals on the ends of the balanced side wires for a neater connection. I sealed the whole package with a silicone sealant. I put a cork in the hole over the toroid but a dab of sealant will also work. I put some sealant where the wires exit the PVC fixture, and around the BNC receptacle flange as well. You can also install a hook if you want to support the antenna from the centre. [*Note:* all photos by the author – *Ed.*]

G4UDU MULTIBAND PORTABLE ANTENNA

Phil Godbold, G4UDU

RadCom 'Antennas' column, March 2013

This antenna was described by Peter Dodd, G3LDO, in his regular 'Antennas' column in RadCom. He wrote: "In last month's 'Antennas' I mentioned that Phil Godbold, G4UDU, is a keen mobile / portable operator and has devised

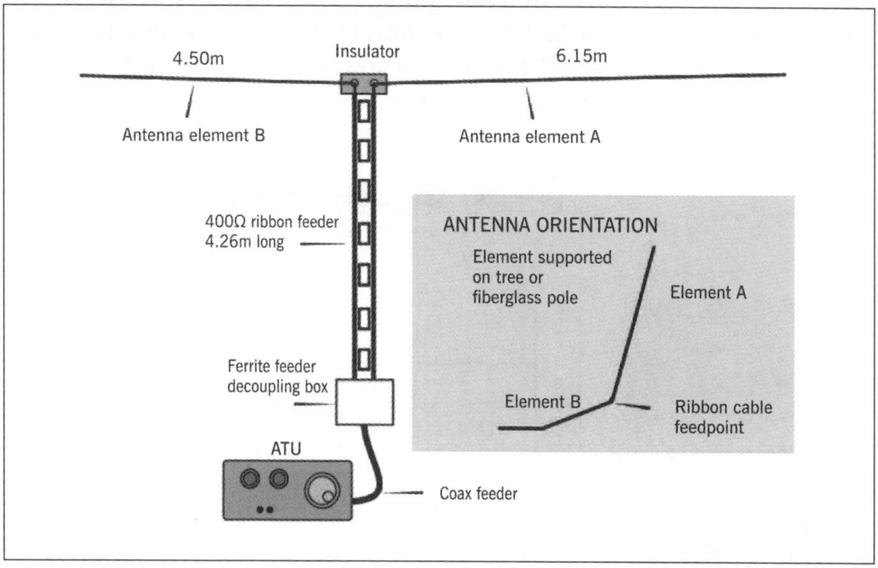

Fig 3.10: The G4UDU portable antenna. The elements are shown horizontally for clarity but are designed to work as a vertical as shown in the inset.

a multiband portable antenna to cover his favourite bands, 7 and 18MHz. It will also work on other bands." This antenna could fit in more than one chapter of this book: it is a vertical but it is also a type of doublet...

The G4UDU multiband portable antenna is similar to the 'Comudipole' [a type of doublet usually used as a fixed station antenna – *Ed*] except that it is smaller, with an overall element length of just 10.65m. The antenna is orientated as a vertical with the top of the vertical part (element A in **Fig 3.10**) fixed to a convenient tree or to a fibreglass fishing pole, with the twin feeder feedpoint about 1.5m above the ground. Element B, the counterpoise, is allowed to lie along the ground, running in the opposite direction to the placement of the feeder.

G4UDU loaned the antenna to Peter Dodd, G3LDO, who wrote: "I set it up as shown in the photo, with the feed point around 4 metres high; this orientation was dictated by the space available at the front of the house. I measured the impedance at the ferrite feeder decoupling box from 6 to 30MHz and the results are shown in **Fig 3.11**."

With the exception of 14.2MHz, the impedances are not all that 'wild' within the amateur bands. Even at 14MHz the impedance becomes more reasonable when the antenna is orientated as designed. The impedance values measured on the amateur radio bands are shown in **Table 3.3**. With the antenna set up as shown in the photo below it loaded quite well using the diminutive MFJ-901B ATU. There was no problem in transmitting but the noise level on receive was so high (S8.6 on 7MHz and S6.2 on 18MHz) that contacts were possible only with very strong stations. This noise level seems to come from the telephone drop wire at this QTH, which is why I locate my permanent HF antennas away from the house. I tried a couple of high impedance current chokes at each end of the coax section of the feeder; this reduced the noise to S6.5 on 7MHz and S4 on 18MHz, which allowed me to test the antenna with DX stations. If you have a receive noise problem at your QTH then common mode (CM) chokes may be the answer.

The G4UDU portable antenna.

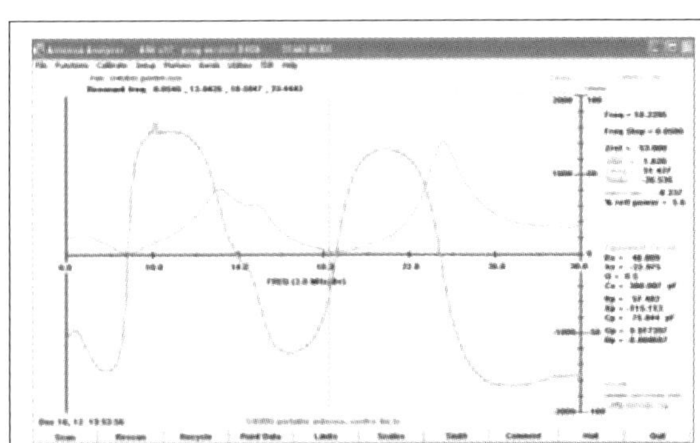

Fig 3.11: Impedance variations from 6 to 30MHz measured at the SO-239 socket of the ferrite feeder-decoupling box.

Freq MHz	Feedpoint 4m high Z	j	Feedpoint 0.5m high Z	j
7.05	105	–203	28	–39
10.1	15	+154	47	+320
14.2	636	–60	216	–340
18.15	47	–30	51	+38
21.2	180	+439	220	+293
24.9	262	–744	262	–609
28.3	60	–337	49	–138

Table 3.3: Measured impedances at the ferrite feeder decoupling box of the G4UDU antenna (converted to the more familiar rectangular coordinates).

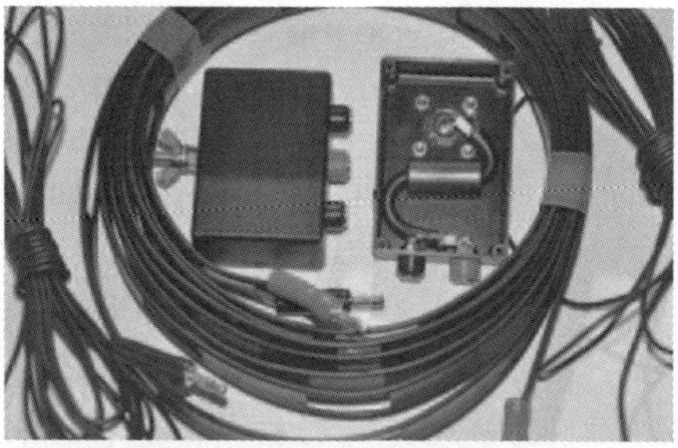

The G4UDU portable antenna with the cover of the ferrite feeder box removed to show the simple RF current choke.

The impedance of the CM ferrite tube located in the ferrite feeder decoupling box is around 200Ω. This is rather low but seems to be enough to reduce CM currents where the antenna is used in a portable location and where noise isn't an issue.

Both transmitting and receiving reports were around 1 to 2 S-points down on short skip and DX contacts compared with my 80ft (24m) end fed on 7MHz and the 30ft (10m) high multiband dipole on the roof on 18MHz; although in some cases the signal reports were the same.

Vertical antennas can be adversely affected by surroundings and they perform best when placed clear of electromagnetic obstacles. When choosing a portable site, selecting a reasonable antenna location can often be easier than from the home QTH.

MODIFYING THE TW2010L BACKPACKERS VERTICAL DIPOLE AERIAL FOR 40M

Nigel Brown, G0GDA

RadCom, September 2015

This antenna project is another that could equally well fit into more than one chapter of this book. Yes, it is a vertical, but it is also a dipole that is being modified. Subtitled "Experiments with antenna construction by an 'amateur' amateur", the author wrote as follows:

Not being a technical person in any way, I have always relied on purchasing my amateur equipment. I spent most of my life as a joiner, finishing up in the maintenance department for the local city council running around in a white van. After moving to my present QTH in a small picturesque village in Kent I thought the residents would probably not have taken kindly to the sight of my tribander and tower appearing overnight. An unobtrusive low profile alternative should fit the bill in the form of a wire, keeping both the neighbours happy and get me on the air again. Following hours of trawling the interweb for information I decided the full-size offset Windom would be ideal, with the parts costing around £80. The construction wasn't as hard as expected, or maybe I was just fortunate. The aerial was put up in the garden at a height of approximately 8m by stringing it between two trees and the cottage chimney. Point to be taken into consideration whilst doing this – the higher one goes up the tree the narrower the trunk becomes. It has proved a more than excellent antenna for the job and using my FTdx3000 I have had contacts from Hawaii in the west to New Zealand in the east using SSB (as CW isn't my forte).

I have always fancied portable work and bolstered with the Windom success I decided to attempt making a 40m dipole by using two legs of 33'8" (10.26m) wire with an 8m fishing pole in the centre for the height, making a V configuration. This was a very simple construction and with the aid of an analyser soon had it working a treat and with my little FT-857D. It is a great portable combination. However, locations can be limited as available posts, trees etc to which the aerial can be temporally attached may not always be available (or be in the wrong spot).

With this in mind I elected to call at my local emporium with a view to buying a go-anywhere, do-anything, standalone system. After a long discussion and copious cups of coffee I came away with a Trans World Antennas TW2010L Backpacker five-band vertical dipole. It is a strange system, looking like a 2-element beam stood on end. But it is a brilliant piece of kit: designed to work on the five bands from 10 to 20m, it can be erected straight out of the bag in about two to three minutes and plonked down on any flattish ground. I tried it out at a local park around 200ft ASL and, using the FT-857D with a spare 12V car battery (which meant the radio output watts would instantly be down). I worked into the Olympic Games Station near Moscow with 59 and a W1 station in Gloucester, MA, with 56 reports.

The original TW2010L antenna.

However, to make the 2010L antenna transmit on the 40m band I found it required an alternative centre control box solely for that band at a cost of $250 [over £190 at the 2020 exchange rate – *Ed*],

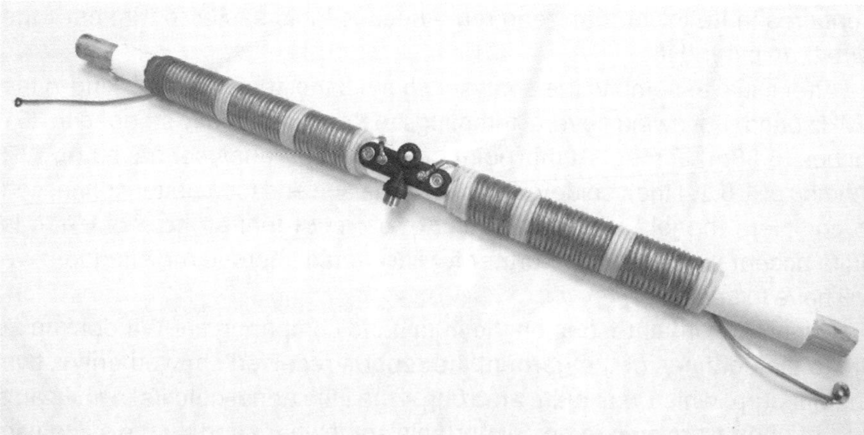

Homebrew centre section for the modified TW2010L antenna.

plus shipping and taxes. So the now-expert constructor – me – rather than spend the pennies decided to make the middle section required.

I purchased the following bits from Wickes: a broom handle and a length of 1.25" plastic waste pipe and, from eBay, the SO-239 fitting. The total came to about £10. I used spare 3mm speaker cable from the shed (I was informed that the thicker the wire the wider the bandwidth) and insulation tape to keep tension on the coils. I used yellow because it's easier on the eye and red so I could tell which end was the top.

A piece of broom handle was cut to the same length as the existing centre section (30.5 inches, 775mm), the idea being to keep the horizontal elements the same distance apart when assembled as per the original. At one end the first two inches was reduced by whittling down the diameter by 1/16" (1.6mm) to make the wood slide into the existing bottom vertical metal tube. The other end of pole fitted directly into the top section with no mod needed. In the top vertical part I drilled a small hole around 3" (76mm) up the tube and fitted a small brass bolt for one of the wire connections – the bottom connection was made using the existing clamp bolt. For a starting point I wound approximately 280 inches (7.1m) of wire on either side of the plastic pipe, as in the photo above. It is better to start with too much cable – which will have the effect of making the

The TW2010L with homebrew centre section.

aerial resonate lower than required – because it is easier to trim back the wire than extend it.

After initial assembly, the analyser showed the aerial was working in the 6MHz band. Following several trimmings of the ends I finally stopped at 251 inches (6.38m) per leg. At this point the MFJ-259B analyser gave an SWR reading of 1.6:1. I then soldered a washer to each end for a better connection to complete the job. Whilst 1.6:1 is not perfect I feel at the moment it is quite acceptable. No doubt further fettling would improve the situation: we will have to see.

During the on-air testing on the 40m band comparing the Windom up at 8m to the confines of the garden, the reports received showed only a two S-point drop, which was quite amazing – the little aerial held its own. I can't wait for the rain to stop to go out portable somewhere and try the system in anger.

I realise only too readily that my simple attempts at construction don't come close to other amateur's projects. I take my hat off to those who can build their own radios and complicated equipment – a talent I could never attain – but that is not my aim as I am writing this article to try and encourage others to have a go at simple construction. Materials needed are usually only a few pounds, tools needed are only basic (or in the case of the analyser, can be borrowed). The main element is time, which is usually free. A little perseverance can go a long way and if for some reason it doesn't work – so what? – all is not lost and at least you have had a go.

I wish you all the best of luck in your experiments, large or small.

The TransWorld Antennas TW2010L Backpacker 5-band antenna is available from the manufacturer in Cookville, Tennessee, USA. See the website at www.mojero.wixsite.com/twantenna — Ed.

Chapter 4

HF Beams

THE VERTICAL DIPOLE ARRAY

David Aslin, G3WGN, and Chris Duckling, G3SVL
RadCom, December 2014

In recent years the Vertical Dipole Array (VDA) has become the antenna system of choice for those portable activities from close to the ocean, especially when the weight of equipment being transported to the portable location is an important consideration. A little history: the first documented use of the VDA was on the February 2008 VP6DX DXpedition to Ducie Island in the Pacific which made an astonishing 183,584 QSOs in 16 days, mainly using VDA antennas. Impressed by the strong signals from VP6DX

A Vertical Dipole Array at TX6G, Austral Islands, 2014.

as received in Europe, Vincent Colombo, F4BKV, designed a series of VDAs based on 10m fibreglass poles for his own one-man operations from several Pacific islands in 2012. Since he had to carry everything himself, weight was of prime concern and each of the resulting F4BKV VDAs weighed only 1.5kg. Vincent described their design and construction on his website **[1]** *and, with his permission, the design was published in the 2014 IOTA Directory* **[2]**.

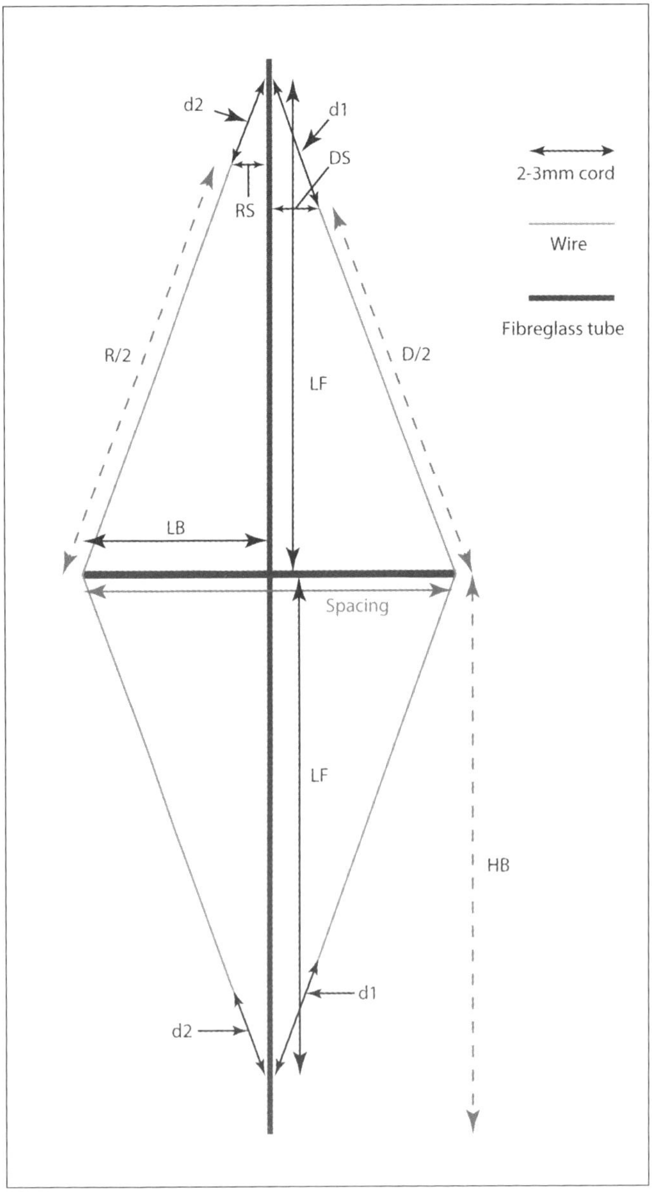

Fig 4.1: The VDA is a 2-element Yagi in the vertical plane with a dipole driven element and a wire reflector behind it.

Vincent's designs were refined further by David Aslin, G3WGN, and Chris Duckling, G3SVL, in preparation for the TX6G operation from the Austral Islands (also in the Pacific) in the Spring of 2014. An article on this DXpedition was published in the June 2014 *RadCom* **[3]**.

Acknowledging the earlier work by the VP6DX team and by F4BKV, G3WGN and G3SVL described both the design and construction of their VDAs. The VDA is a 2-element Yagi in the vertical plane with a dipole driven element and a wire reflector behind it. The basic design of the VDA is shown in **Fig 4.1**. The ends of the elements are brought close together to form a diamond shape .

This results in a relatively broad beam in the horizontal plane, about 139° at the 3dB points, as shown in **Fig 4.2**, but with an important lowering of the elevation angle at which maximum radiation takes place, to just 7° above horizontal. It is this low radiation angle on a site of high ground conductivity with a take-off over an ocean path that combines to produce the 8dB or more gain reported relative to a horizontal dipole one half-wavelength high over good ground.

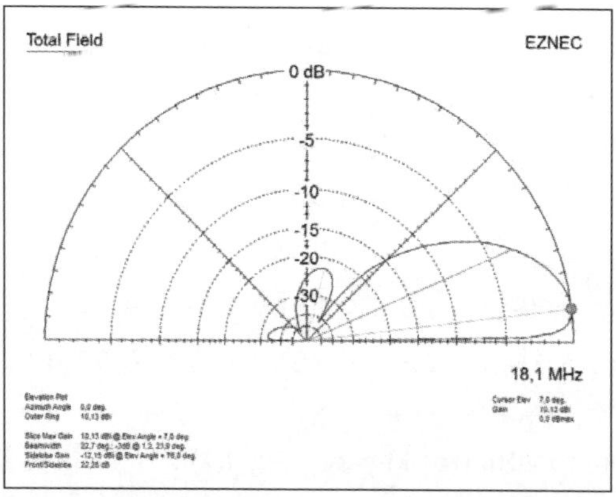

Fig 4.2: The VDA has a relatively broad beam in the horizontal plane; this is the elevation plot.

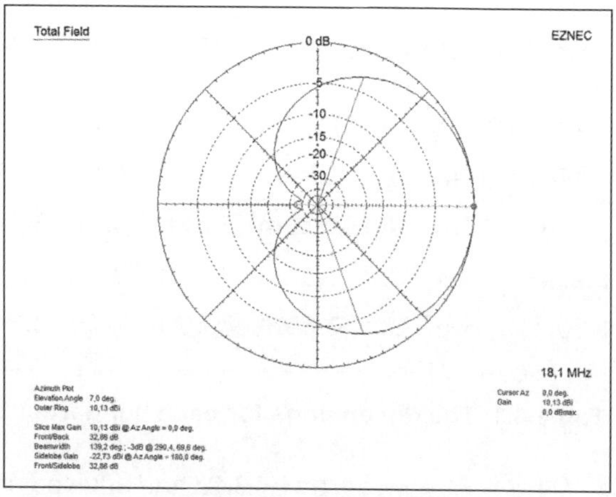

Fig 4.3: The design has an approximately 30dB null off the back of the beam; this is the azimuth plot.

As seen in **Figs 4.2** and **4.3**, the VDA has an approximately 30dB null off the back of the beam.

A note of caution: TX6G, VP6DX and other stations using VDAs had strong signals because of the combination of low-angle antennas with a location on the beach, usually within a quarter-wavelength of the high tide mark. While a VDA will work inland (such as in your garden) and will have directivity and gain over a vertical dipole or quarter-wave ground plane, its take-off angle will not be anything like as low as 7° and its efficiency will be much reduced when compared with the VDA on a beach. The proximity to the ocean matters! But, if a vertical is all you have room for, by all means give the VDA a try, it may be better than what you have. And if you are operating portable from a beach, whether in the UK or at a holiday location overseas, the VDA is an excellent choice to take with you.

With F4BKV's work as inspiration, G3WGN and G3SVL set about adapting the mechanical design to materials easily available in the UK. They had some straightforward design goals:

- Simple assembly / disassembly
- Easy adjustment to the desired part of the band
- Easily obtainable parts (from DIY stores or online)
- Lightweight
- Disassembled length suitable for all airlines
- Easily water- and salt-proofed
- Delivers gain relative to a dipole in the same location

G3WGN and G3SVL decided to build all the VDAs for the five bands 10m to 20m and test them both mechanically and electrically before their TX6G operation. "We also wanted to investigate how critical the dimensions

Band	Mast	Boom	LB	HB	LF	D/2	R/2	DS	RS	d1	d2
20m	12.5m	4m	1.78m	6.24m	6.0m	4.77m	5.07m	0.46m	0.42m	1.62m	1.51m
17m	10m	3m	1.39m	4.87m	4.6m	3.69m	3.97m	0.37m	0.33m	1.30m	1.19m
15m	10m	3m	1.19m	4.45m	4.2m	3.18m	3.40m	0.31m	0.28m	1.09m	1.01m
12m	8m	3m	1.01m	3.91m	3.7m	2.70m	2.88m	0.26m	0.24m	0.91m	0.87m
10m	8m	3m	0.89m	3.66m	3.4m	2.39m	2.55m	0.23m	0.19m	0.81m	0.67m

Table 4.1: The dimensions for each band from 10m to 20m (for key see Fig 4.1) .

were, as F4BKV had advised caution in some areas. This turned out to be a valuable exercise and undoubtedly saved us much time and angst when setting up at the DXpedition site," they wrote.

The dimensions for each band from 10m to 20m are shown in **Table 4.1**. These are different from those on F4BKV's website as they were recalculated based on the wire used for the elements (DX-Wire Ultralight). These dimensions were found to be a good starting point and produced antennas that required only a moderate amount of 'tweaking' to get to the right frequency. They should also work for typical wire you may have to hand; just be sure to have sufficient excess when first making the elements, to allow for later adjustments. It was found that the dimensions from the pre-DXpedition trials at home gave similar resonances on the beach at TX6G.

The lengths of the driven element and reflector may look odd, appearing to be far too short, but be assured these are the correct dimensions and they will work. The shortening is due to the effect of the geometry of the VDA, especially the angle of the elements and the closeness of the element ends to each other.

The design could, of course, be scaled up to 30m or even 40m but the mechanical challenges of a 25m mast with an 8m boom were more than could be accommodated within the limited weight and size constraints (elevated quarter-wave ground planes were used on those two bands instead.)

The basic supports could be fishing poles, but good quality vertical supports such as those from Spiderbeam **[4]** or DX-Wire **[5]** are recommended. The booms, however, can be made of lightweight fishing poles.

The centre boom-to-mast plate needs to be strong enough to support the boom as the antenna is raised and to keep the boom reasonably horizontal once up in the air. G3WGN and G3SVL ended up using £1 supermarket 'Value' range chopping boards with a length of 25 x 16mm electrical conduit as a runner into which the boom fishing poles are clamped, as shown in the photograph at the top of the following page. The clamps are made of tie wraps slipped through rubber sleeving that permits the tie wraps to be pulled tight without crushing the fishing rods. This form of fixing was also used to secure the centre boom-to-mast plate to the vertical mast.

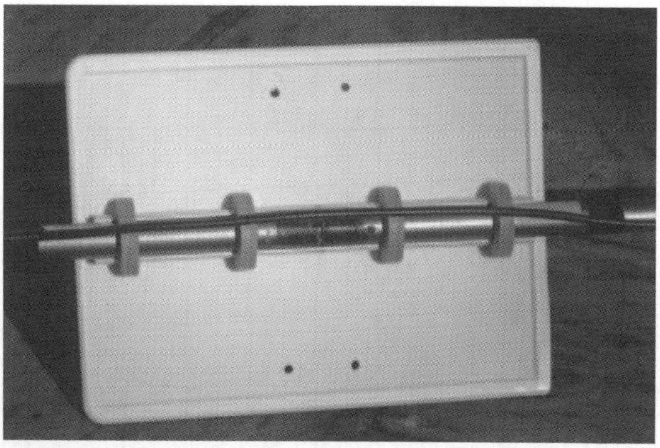

The boom-to-mast plate.

Commercial dipole centres were used for the driven element following problems with salt ingress on the electrical terminal blocks used on a previous operation. No problems were encountered during the two-week activity. TX6G did, however, experience arcing at the tips of the elements once salt from the sea spray had coated the end insulators. It is an easy matter to wash the bottom insulators when necessary, but reaching the top ones required the antenna to be lowered. One solution would be to double insulate the ends of each wire with heat-shrink tubing.

A simple ferrite choke using a Fair Rite FT240-31 toroid was used on the coax at the feedpoint to maintain the radiation pattern and to reduce common mode currents on the outer of the coax. The ferrite's weight is supported by passing it over the boom and securing it with tie wraps as shown in the photo below. The boom also acts as a support for the coax, which should be run away from the reflector and kept horizontal for as far as possible behind the antenna. There is no need for the boom to be cut to size: an overhang at the driven element is of no consequence and an overhang behind the reflector is useful to keep the coax horizontal as it leaves the array.

When assembling the VDA, G3WGN and G3SVL found the design to be very tolerant with no really critical parameters. There are, however, a few important points to note. The most critical dimensions are the reflector spacing and driven element spacing between the ends of the wires and the centre line of the mast – they should be within a centimetre or two of the design distances. If not, adjust the length of the support cords or tension them until they are.

Guys should be attached about 30 – 50cm above the boom to stabilise the mast. A guy ring should be used if it is intended to rotate the array, although in many cases this may not be necessary as the front lobe is wide.

To tune the VDA, attach an antenna analyser to the feeder and plot the SWR curve. Adjust the length of the

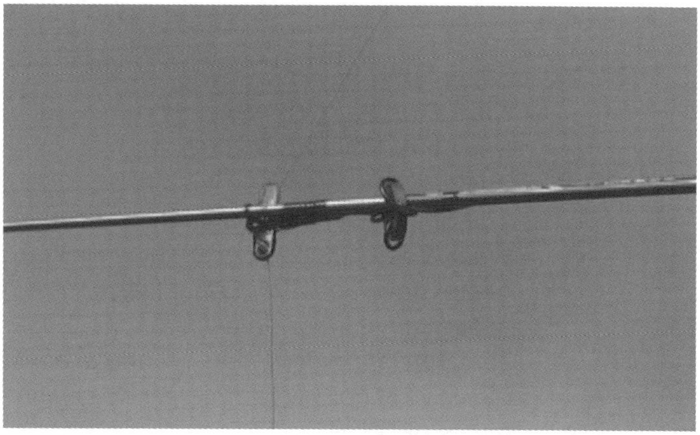

The dipole centre and ferrite choke balun attached to the boom.

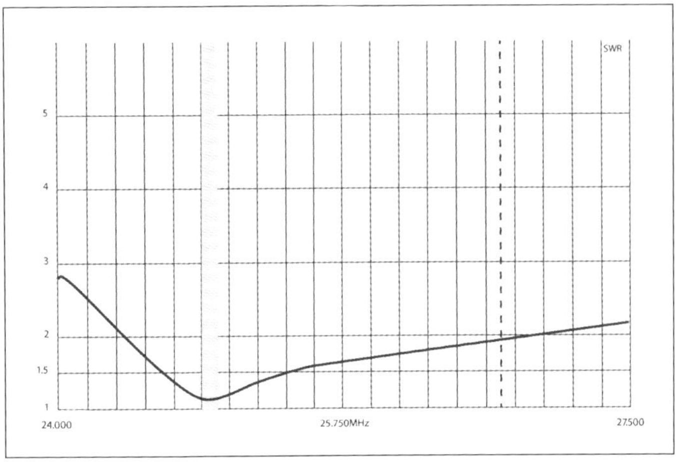

Fig 4.4: The SWR curve for the finished 12m VDA.

lower end of the driven element to put the array at the desired resonant frequency. Fold back and wrap the tail around the element. Only the lower end of the driven element needs be adjusted as small changes are all that are required and the array is inherently asymmetrical; there is no need to shorten (or lengthen) both ends of the driven element as one would a horizontal dipole.

Repeat the adjustment until the SWR curve is optimised. If your adjustment of the driven element requires lengthening or shortening by more than, say, 4cm, it is wise to adjust the reflector by the same amount to keep the relative driven and reflector lengths in approximately the same ratio. This should minimise any change in antenna pattern and gain during adjustment. However, the reflector adjustment interacts with the driven adjustment, so it may take several attempts until you have an SWR curve with a minimum at the chosen frequency and elements in a similar ratio to that given in the table. The SWR curve for the finished 12m VDA is shown in **Fig 4.4**.

G3WGN and G3SVL concluded by saying that while VDAs are not 'magic' they do provide a low cost, low weight, easy-to-build solution for beach-located portable operation. "They provide gain, a reasonable front-to-back and a broad forward lobe of radiation and most important, provided us with many unsolicited 'you are loud!' comments. Our experience and that of F4BKV and [others] before us shows they are reproducible and effective."

THE SPIDERBEAM

Cornelius Paul, DF4SA
RadCom, November 2003

The Spiderbeam is a well-known commercial antenna made in Germany. What is perhaps not so well known is that it (and the company that bears its name) developed from what was originally a home-brew project designed specifically for portable operations. While there are now eight varieties of commercial Spiderbeam antenna available (plus 40m dipole add-on kits), the original home-brew project was a light-weight beam covering the three bands of 10, 15 and 20m.

The November 2003 *RadCom* article was a construction project, but the original design has been developed and improved over the intervening years.

Rather than re-publish the 2003 constructional details, if you wish to build one of these antennas from scratch take a look at the Spiderbeam website **[4]** which has full construction guides (in numerous languages) that can be downloaded as PDFs. Nevertheless, Cornelius Paul's original article still makes for interesting reading and gives an insight into the requirements that led to him developing the antenna as well as a useful overview of its design and construction.

He wrote, "My favourite way to play radio is on portable, outdoor operations, field days, and DXpeditions. I definitely needed a lightweight antenna, and the Spiderbeam was developed to fulfil that need. It yielded the following benefits: it should be lightweight (5.5kg) and have a small packing size (1.20m). These make transportation a lot easier, even more so because a light mast and rotator are sufficient. The antenna also presents low wind resistance."

The Spiderbeam he described was a triband Yagi for 10, 15 and 20m, consisting of three interlaced wire Yagi antennas strung on a common fibreglass 'spider'. These are three-element Yagis for 20m and 15m, and a four-element Yagi for 10m. In contrast to a regular Yagi, the director and reflector elements are V-shaped (see **Fig 4.5**). The driven element is a multi-band fan dipole for 20, 15 and 10m, i.e. three individual dipoles connected at their centre feed point. The feed point impedance is 50Ω, fed through a W2DU-type current choke balun, making a very simple and robust feed system. No phasing lines or matching devices are needed.

The heart of the mechanical construction is the central hub made from aluminium sheet metal and tubing. Most antennas' boom-to-mast plates put the antenna on one side of the mast but with the Spiderbeam's central hub the mast goes right through the centre of mass. The antenna's weight and vertical torque momentum are optimally applied to the mast and rotator, which means the load on these parts is reduced.

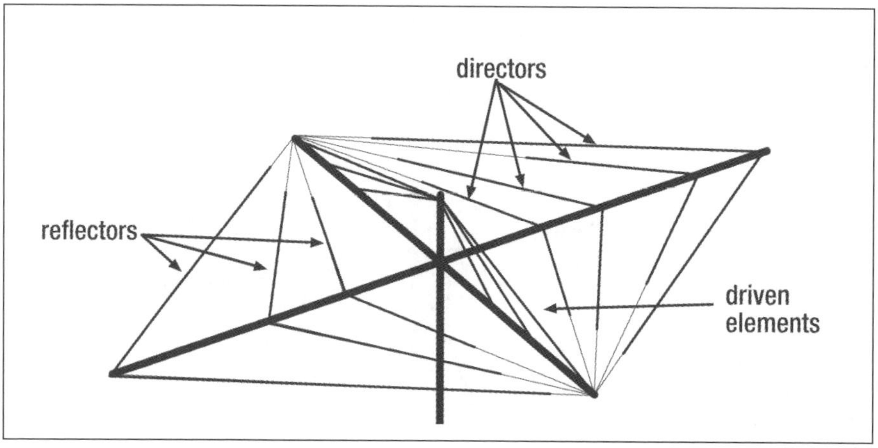

Fig 4.5: Layout of the original 10-15-20m homebrew Spiderbeam project, showing fan dipole driven elements and V-shaped reflectors and directors.

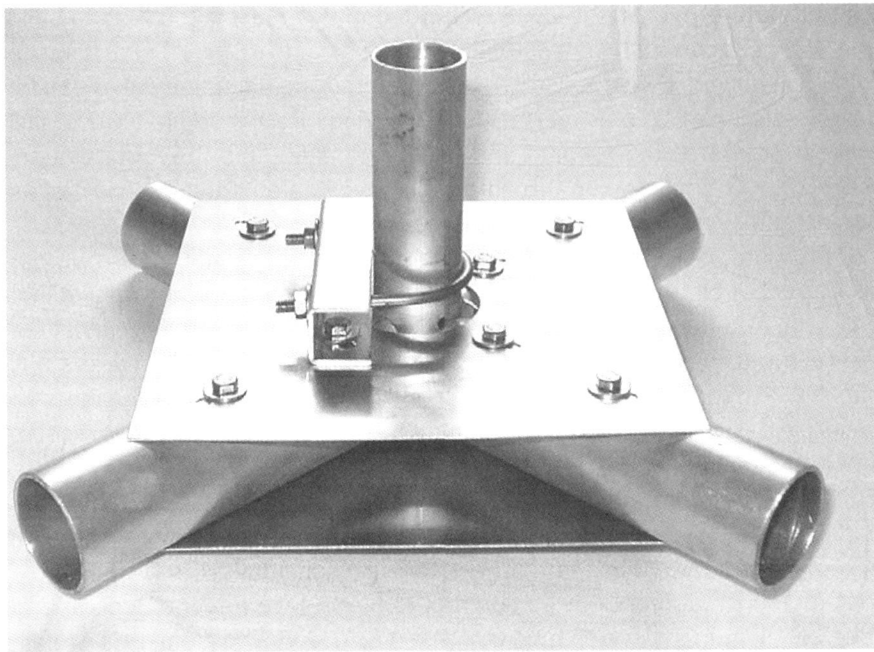

The central hub into which the four fibreglass spreaders are inserted. The vertical antenna support mast goes right through the centre of the Spiderbeam, thus distributing the load evenly.

In the original design, the fibreglass tubes were the bottom 5m-long sections of 9m-long fishing rods, making the antenna 10m x 10m in size. The 'spider' (**Fig. 4.6**) gets its extra stability by guying it completely within itself: the guys are Kevlar lines (1.5mm diameter, 150kg breaking strength). Kevlar has the big advantage of not stretching at all, so the guy-lines always stay as tight as you pulled them during assembly.

DF4SA pointed out how critical the wire lengths are. The lengths he quoted were valid only for bare wire of 1mm diameter. Other wire types

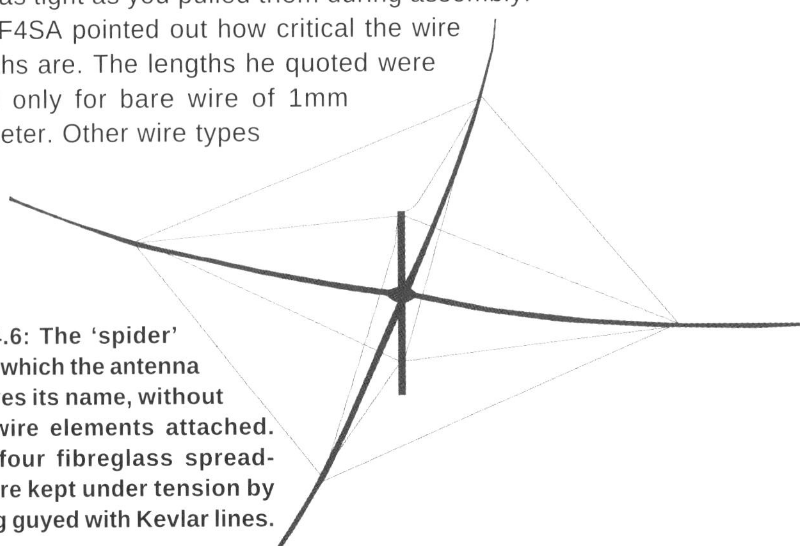

Fig 4.6: The 'spider' from which the antenna derives its name, without the wire elements attached. The four fibreglass spreaders are kept under tension by being guyed with Kevlar lines.

Operating frequencies	14.0 – 14.35 MHz
	21.0 – 21.45 MHz
	28.0 – 28.8 MHz
Feeding	Single 50Ω coaxial cable
Weight	5.5kg
Turning radius	5.0m
Transportation length	1.2m
Rotator requirement	TV rotator

Band	Forward gain in free space	F/S ratio (dB)	F/B ratio (dB)
20m	6.5 dBi (4.3 dBd)	12	15–20 across band
15m	6.6 dBi (4.4 dBd)	15	18–25 across band
10m	7.2 dBi (5.0 dBd)	18	20–30 across band

Table 4.2: Mechanical and electrical specifications of the original 10-15-20m lightweight Spiderbeam.

(especially insulated wire) would result in different element lengths because of the change of velocity factor caused by the insulation. The same is true when mounting insulators at the wire ends, as they will also cause a change of the effective electrical wire length. Even an error of one centimetre will make a difference. It is also essential to use a wire type which does not stretch. He used copper-clad steel wire (Copperweld®, DX-wire®) although the earlier prototypes were built using normal (soft) enamelled copper wire. Each time when assembling and dismantling the antenna, DF4SA found that some elements had stretched up to 10cm. In turn, the resonant frequencies of the wire elements changed, resulting in a deterioration of the radiation pattern, especially the front-to-back ratio. Attaching the wire elements to the spider is quick and easy, using cable ties. Short pieces of polyamide hose were used for stress relief at the bending points and also to cover the joints between the wire ends and the fishing line insulators.

The electrical and mechanical specifications of the original 10-15-20m lightweight Spiderbeam are shown in **Table 4.2.**

Because the feedpoint impedance of the antenna is already very close to 50Ω no impedance transformation is necessary, though a balun between the antenna and the coaxial cable is needed. This could take the form of a simple coaxial cable choke, constructed by coiling up five to 10 turns of coaxial cable at the feed point. However, the performance of such a choke is highly dependent on the operating frequency, the coaxial cable

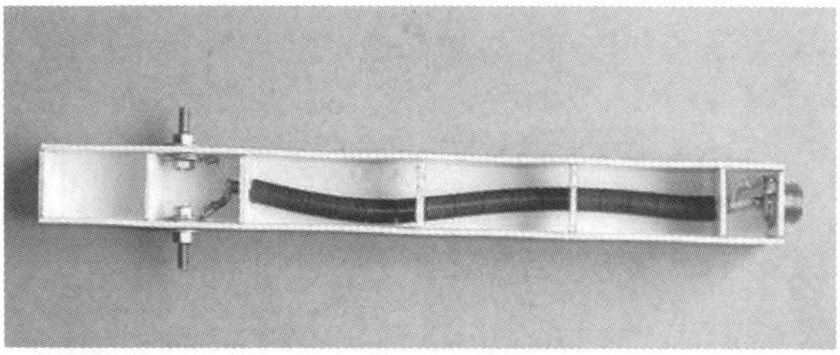

The W2DU-type coaxial cable choke balun as used in the origial Spiderbeam design.

used, and the diameter and height of the coil. A much better solution is the coaxial choke developed by W2DU: take a piece of thin coaxial cable and slip a number of ferrite beads over the outer plastic jacket, which effectively stops current from flowing on the braid (outer conductor), resulting in a good match of the balanced antenna to the unbalanced coaxial cable. Using a piece of Teflon cable makes such a coaxial cable choke easily capable of handling 2kW continuous RF power.

To summarise, assembly of the Spiderbeam is straightforward, user-friendly and, other than the wire lengths, non-critical (the spacing of the elements is not critical). No complicated or fragile parts are involved and the tuning procedure requires only an antenna analyser or SWR bridge and can be done in 10 minutes.

For transportation, all the wires and guy-lines are wound on to a big spool in the most appropriate order for assembly.

The commercially-made Spiderbeam antennas are described in the chapter on commercial portable antennas.

THE HEXBEAM

Peter Dodd, G3LDO
RadCom 'Antennas' column, January 2008

The Hexbeam, like the Spiderbeam, is now a commercially-available antenna and it also started life out as a home-brew project. The design was so successful that there are now several manufacturers around the world building and selling

The multiband Hexbeam constructed by Steve Hunt, G3TXQ.

them. Being relatively small (for an HF beam), and very light in weight, it is ideal for portable operations where gain over a dipole or vertical is desired.

The Hexbeam is also a wire beam built on fibreglass spreaders, but it is very different from the Spiderbeam in several ways. It has only two elements per band (a driven element and a reflector) and although it also uses full-size elements, the wires are bent into 'M' or 'W' shapes which reduces the overall size of the antenna considerably.

The Hexbeam could therefore be thought of as a 'mini-beam' but, since it uses full-size elements, there is only a little compromise on gain. The original design had good gain and front-to-back figures but at the expense of a very narrow operating bandwidth.

This is where the late Steve Hunt, G3TXQ, came into the picture. He contacted the *RadCom* 'Antennas' columnist, Peter Dodd, G3LDO (also Silent Key), who had been writing about compact HF beams in the column. In the *RadCom* 'Antennas' column G3LDO quoted Steve Hunt as saying, "I've spent recent weeks doing more Hexbeam experiments on a 10m test bed antenna. In particular I've looked at a method of extending the bandwidth, searched for the optimum driver / reflector tip spacing, and found a suitable driver / reflector ratio that lends itself to Beta matching."

The Hexbeam design is shown in **Fig 4.7**. It is constructed of wire elements strung on to a support structure of fibreglass spreaders. By 'bowing' the spreaders upwards, as can be seen in the photograph of G3TXQ's

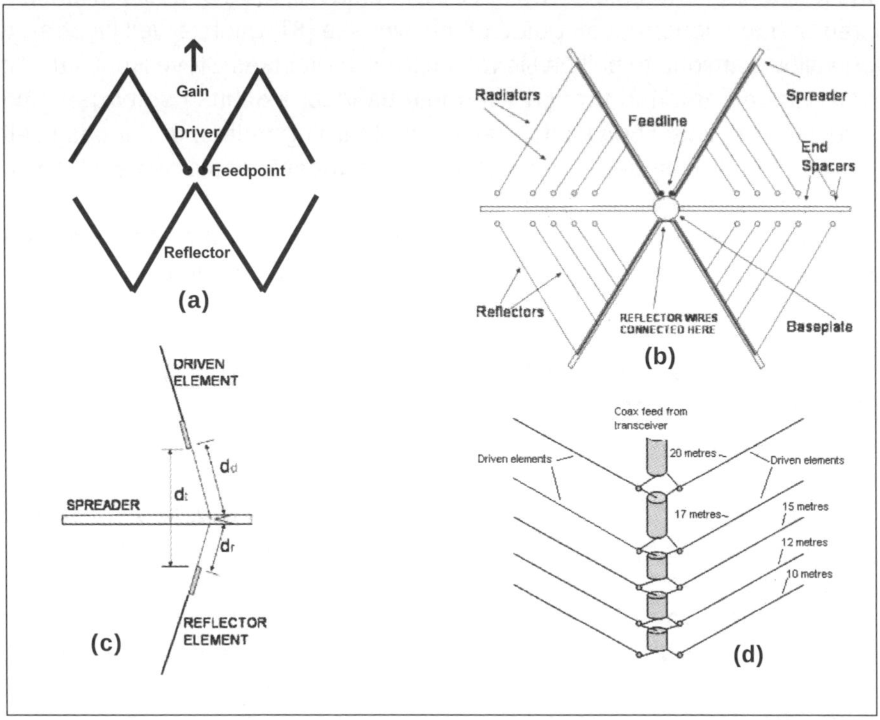

Fig 4.7: Various aspects of the design of the Hexbeam.

Practical implementation of the feed method, by K4KIO.

Hexbeam, the elements for several bands can be accommodated on the same support structure and spaced apart in both the horizontal and vertical planes. This approach allows a multiband beam to be constructed with a turning radius about half that of a corresponding Yagi. Peter Dodd commented, "The method of feeding [shown in **Fig 4.7(d)**] is interesting... When you hear about the problems of feeding multiband quads it comes as a surprise that the method of feeding the multiband Hexbeams is so simple." The practical implementation of the feed method is illustrated in the photograph to the left.

In another e-mail to G3LDO, G3TXQ said that "Leo Shoemaker, K4KIO, and I have cooperated on the [Hexbeam]... we agreed that I would major on the theory and experimental work, while he covered practical constructional information... We decided to try to produce a definitive set of dimensions. The recommended dimensions on his site were a joint effort; they are well proven and are the result of many hours of *EZNEC* simulation followed by the construction of at least six 'real' antennas."

Once again, the design of what has now become known as the 'G3TXQ Broadband Hexbeam' has undergone some changes since the 2008 correspondence between G3TXQ and G3LDO, in particular to the shape of the reflector elements. However, Leo Shoemaker, K4KIO, has published a step-by-step construction guide on his website **[6]** which is well illustrated and allows anyone to build a Hexbeam from scratch. As Steve Hunt said, "If you want an antenna that can be multi-banded, exhibits useful gain and directivity, is very lightweight, has a small turning radius, and lends itself readily to 'do-it-yourself' construction, the Hexbeam should be very high on your list of options".

A description of one of the commercially-made Hexbeams on the market can be found in the chapter on commercial portable antennas.

A PORTABLE TWO-ELEMENT 15-METRE YAGI

Jack B Morgan, KF6T
QST, July 2008

This inexpensive, lightweight and quick-to-assemble Yagi for 21MHz was designed and built by Jack, KF6T, especially for ARRL Field Day operations but it would be equally suitable for any portable activity.

ARRL Field Day 2007 was on the horizon and our local club did not have an antenna lined up for 15m. We were worried that the band might open up during the event and important opportunities for contacts would be missed. It happened in 2006, and only a few contacts were made using our 40m antenna, so we needed a better solution.

Some of the requirements established were that this antenna be compact, light weight, easily assembled, no fancy matching required and it needed to cover the entire band.

I selected a two-element design to meet these requirements. I have had experience with this approach in the past and have built similar antennas from 40 to 6 metres with good results. With a small sacrifice in gain and front-to-back ratio (F/B), you can always achieve a direct 50Ω match with two elements on a short boom. I used *EZNEC* to confirm the element dimensions and spacing.

The elements are isolated from their supports so the clamps have little effect on dimensions or performance. I was so confident that the computer-generated dimensions were correct that I didn't test the antenna until we raised it on Field Day weekend. I was relieved that the VSWR was exactly as predicted (see **Fig 4.8**). The new antenna played very well during the

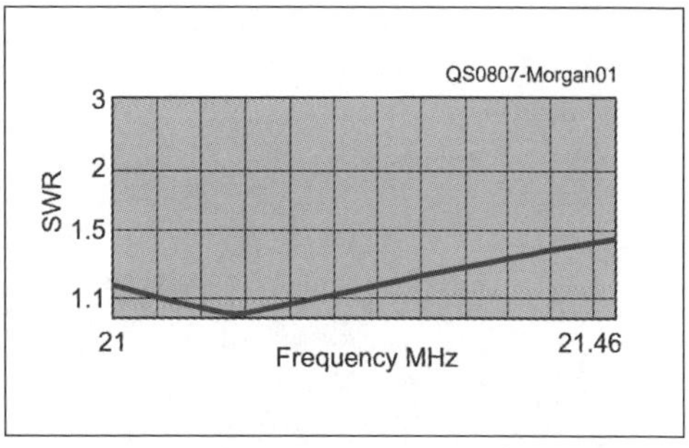

Fig 4.8: SWR curve of the 15m portable 2-element Yagi antenna.

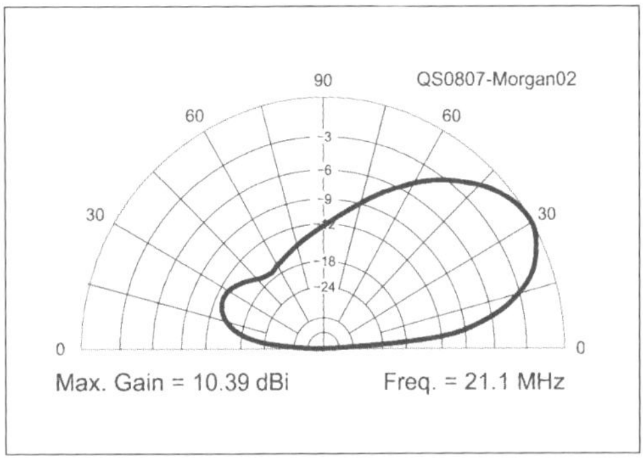

Fig 4.9: Elevation plot of the 15m antenna at 20ft, in the peak of the main beam.

15m openings. The main lobe of propagation is at 30° elevation at 20 feet above flat land (see **Fig 4.9**) and even lower at our cliff-like location. This provides good propagation over the intermediate distances important to Field Day.

The most unusual aspect of this antenna is that it uses a boom with a square cross section. I used an 8 foot piece of 1.25 x 1.25 inch 6061-T6 aluminium with 1/8 inch wall thickness. This boom is very strong and offers some interesting advantages. First of all, no U bolts are required except for mast attachment. Everything else just bolts together. Secondly, because the mounting surfaces are square, element to mast alignment is guaranteed! We assembled this antenna in 10 minutes at the Field Day site. All components are 8 feet or less, so this antenna is easy to transport or store.

The two elements are made from 5/8 (0.625) and 1/2 (0.500) inch 6061-T6 aluminium tubing with 0.058 inch wall thickness. These tubes will usually telescope together but sometimes need a little adjustment with fine sandpaper. If the fit is too tight, you can get into a situation in which the elements will gall and disassembly will be difficult. If you use any sort of lubricant, apply it just before assembly to keep it from acting like fly paper to dirt or sand.

The elements are connected together at a 2 inch overlap point using a single 8-32 stainless steel screw and Nyloc lock nut.

Oatey 33940 1/2 inch EMT clamp used to secure elements.

The two element clamps are fabricated from a one foot section of 3 x 1 inch aluminium U channel with 0.125 inch wall thickness. Note that the boom attachment holes are initially made smaller as pilot holes. At final assembly the finish drilling is accomplished.

I used grey polypropylene conduit clamps as shown in the photographs to support the elements on the boom. These clamps are intended to be used to mount 1/2 inch electrical conduit to a wall. They

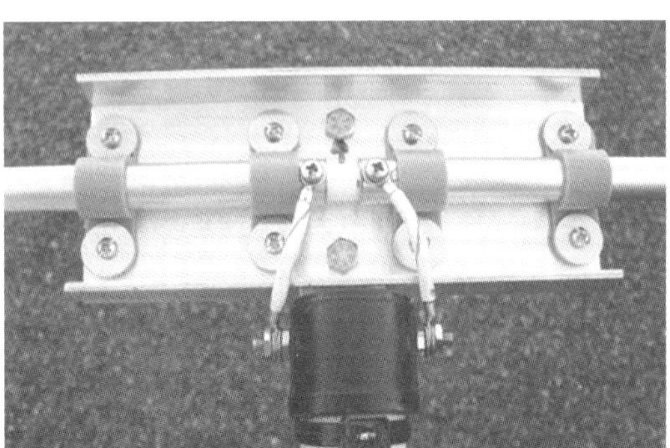

Driven element assembly.

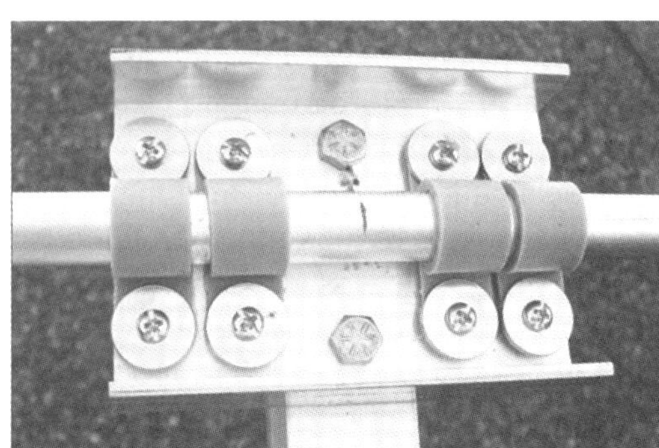

Reflector assembly.

Item	Overall Length	Cut Length	Notes
Driven element	259.25		Check tip to tip dimensions.
2 x 5/8" inner sections	47.75	47.75	Cut from one 8' section.
2 x 1/2" element tips	81.625	83.625	Allows 2" overlap.
Reflector	280.75		Check tip to tip dimensions.
1 x 5/8" centre section	96.0	Not cut	Check stock is 96" long.
2 x 1/2" element tips	92.375	94.375	Allows 2" overlap.
Boom	96.0	Not cut	Use 8' section of 1.25 x 1.25"
Driven element to reflector spacing	93.0		Clamps flush at boom ends

Table 4.2: 15m portable Yagi dimensions (inches).

work very well in this application. Fender washers (0.75 inch diameter) are used under the 8-32 screws to support the poly clamps better. The U channel plates are tapped to accept these screws and are secured on the far side with stainless steel lock nuts.

The element dimensions are shown in **Table 4.2**. The driven element has a half inch wide gap that is filled with a half inch diameter insulating rod made of fibreglass, Teflon or Delrin. The 2.5 inch long insulator is inserted 1 inch into each split elements. I drilled and tapped (8-32) holes through the tubing wall and into the centre of the insulator.

By not drilling all the way through, we prevent an oversized screw from shorting on the far side at the point of balun attachment later. It is a lot easier to drill these holes with the elements and insulator mounted on the support clamp beforehand.

Each element clamp is mounted to the boom using two 1/4-28 (UNF fine thread) stainless bolts. I clamped the parts together, got them square, then drilled (0.250 diameter) through the clamp pilot holes continuing through the boom in two places. Mark the mating parts with a permanent marker pen so they will exactly match up at a later time.

A current balun is used to feed the driven element. This keeps the coax from radiating unwanted common mode signals, an important consideration in a Field Day environment. It is mounted as close as possible to the driven element and held in place with tie wraps.

The mast to boom plate is fabricated from a scrap piece of 0.25in thick aluminium plate. The U bolt hole dimensions will depend on the actual mast diameter selected. This design used 2-inch U bolts and is shown mounted on the square boom in the photo. I

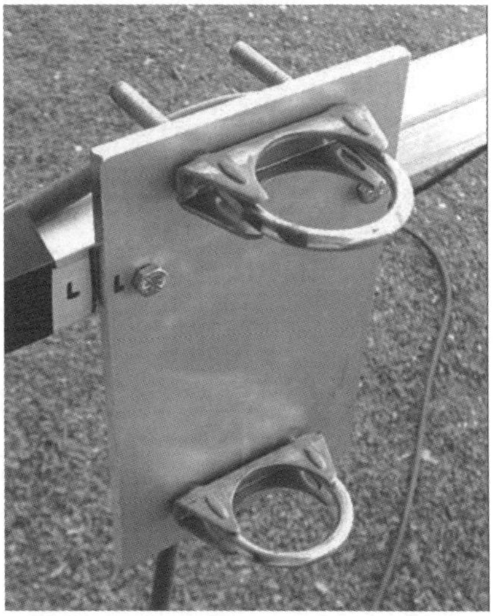

The mast to boom plate installed on the square boom.

located this plate at the centre of balance of the antenna, with the balun mounted.

The completed antenna was mounted on a windy ridge top 7000 feet up in the Sierra Nevada Mountains during the 2007 ARRL Field Day. The small diameter elements shrugged off the wind and displayed an acceptable amount of droop. Analysis using Dave Leeson's spreadsheet program (element strength calculator by W6QHS / W6NL) shows that these elements should survive 72mph winds.

REFERENCES

[1] F4BKV VDA antennas: f4bkv.net/antenna-vda.html
[2] 'Antennas for IOTA DXpeditions', Steve Telenius-Lowe, PJ4DX, *IOTA Directory*, RSGB 2014.
[3] 'TX6G – Raivavae – Austral Islands', Don Beattie, G3BJ, *RadCom* June 2014, pp57-60.
[4] www.spiderbeam.com
[5] www.dx-wire.de/lng/en/fiberglass-poles/
[6] Hexbeam construction article by Leo Shoemaker, K4KIO: hex-beam.com
[7] *EZNEC*: www.eznec.com

Chapter 5

VHF and UHF Antennas

A FOLDING 2m YAGI

Eric Lake, 2W0WXM
RadCom, January 2013

Eric Lake, 2W0WXM, shows how a little ingenuity can provide an antenna that is perfect for transporting to a local hilltop for a quick portable operating session.

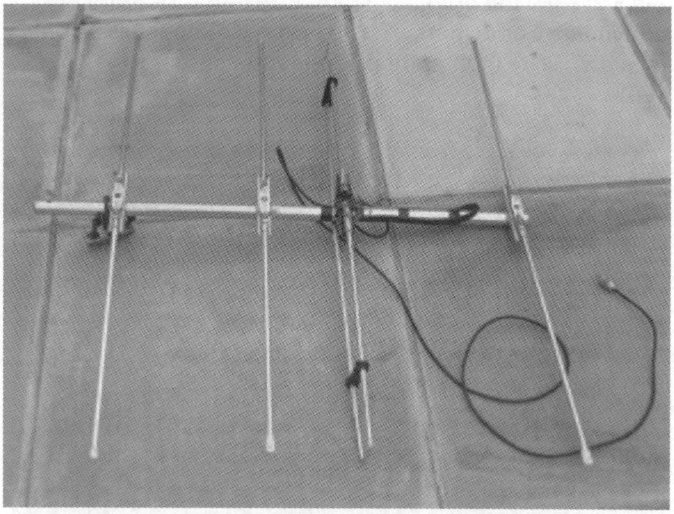

The foldable antenna fully open and ready to go.

Eric wrote that he saw an old 4-element Yagi for the FM broadcast band for sale for £1 at a car boot sale and wondered if he could convert it into a 2m beam for portable operations.

The antenna had all its elements fixed in place with wing nuts. It was folded up and standing vertical, only taking up about 15cm square except for the driven element, which was a separate item. The basic construction of the antenna was a boom with three metal cross-pieces to which the reflector and director elements were attached using wing nuts. A plastic insulator provided support for the folded driven element. All of the elements were made from aluminium that had been rolled into a tube shape but not welded. The ends of the reflector and directors were flattened and rounded, presumably for safety reasons.

Eric realised that he could quite easily shorten the elements and adjust the spacings to make the antenna work on 2m. **Fig 5.1** shows the general arrangement of

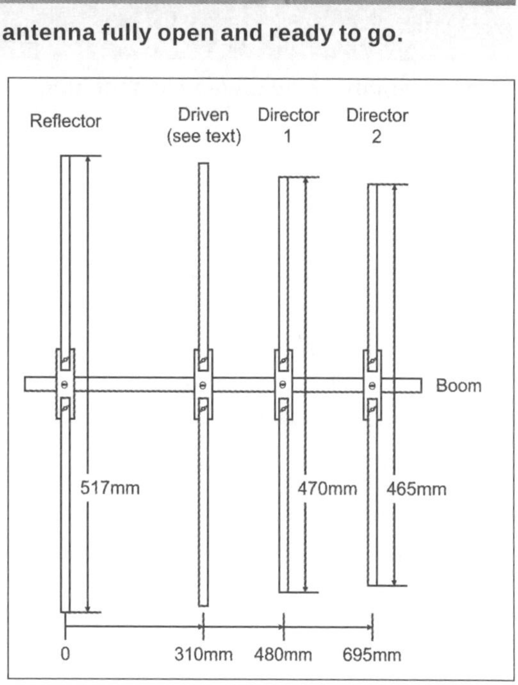

Fig 5.1: Main dimensions of the foldable antenna.

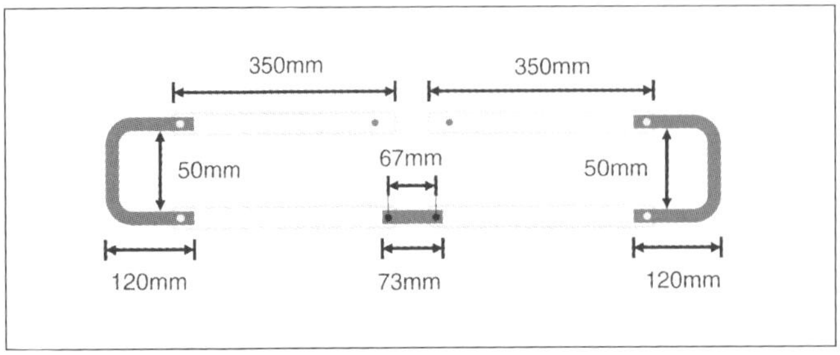

Fig 5.2: How the driven element was constructed. The aluminium bar is in black, the original element material is in grey. The feed point is at the top.

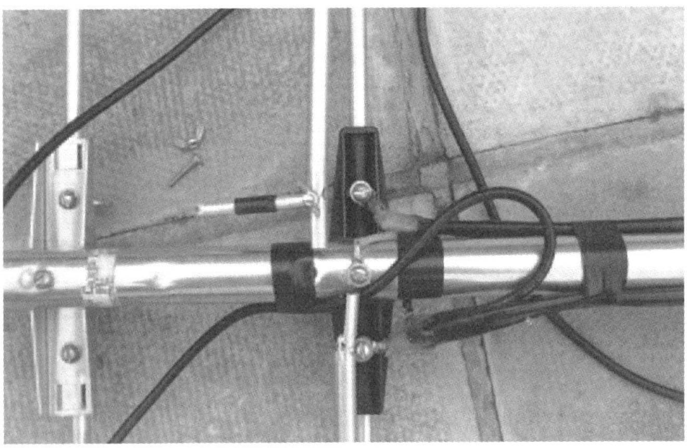

Final assembly of the driven element hinge.

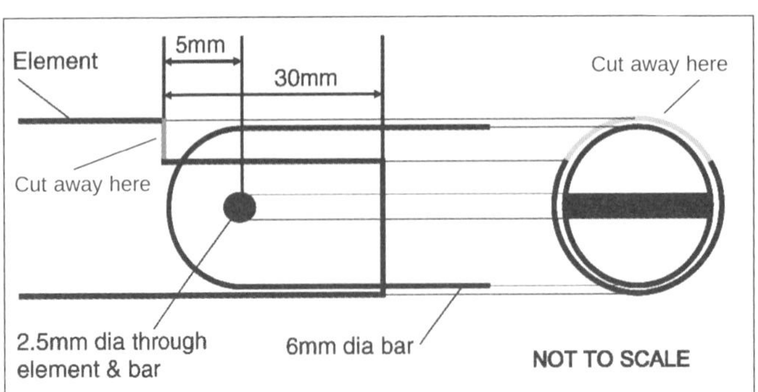

Fig 5.3: Final dimensions and drilling details of the driven element bar.

the revised antenna elements and their spacings. The antenna was disassembled and suitable holes drilled in the boom for the elements at the new spacings. The reflector and director elements were shortened to the new lengths, with new holes drilled for the wing nuts.

The key to making the whole antenna foldable was to work out a way of hinging the driven element. Given that, like the other elements, the two ends of the driven element were already attached using bolts and wing nuts, it seemed sensible to work out a way of making the middle of the 'solid' side fold. Eric did this by cutting out a section of the tube and replacing it with a short section of 7mm solid aluminium bar. The size was chosen so it would slide inside the driven element tube. The fixing centres for the bar are the same as those for the 'open' ends of the driven element, 67mm. **Fig 5.2** shows the eventual dimensions and the photo shows the final assembly. A file was used to round off the ends of the bar.

To allow for the hinge bar to fold at 90° it is necessary to cut out a section of the element tube, as shown in **Fig 5.3**. Prise the slot open just enough to be able to open up the tubing to allow the hinge bar to reach the 90° angle needed for it to fold.

The next thing to work on was the length of the driven element. Eric wrote, "In order to make life easy for myself I decided to cut off the existing folds of the driven element and replace them with a section of solid aluminium bar that I folded into a U shape." The

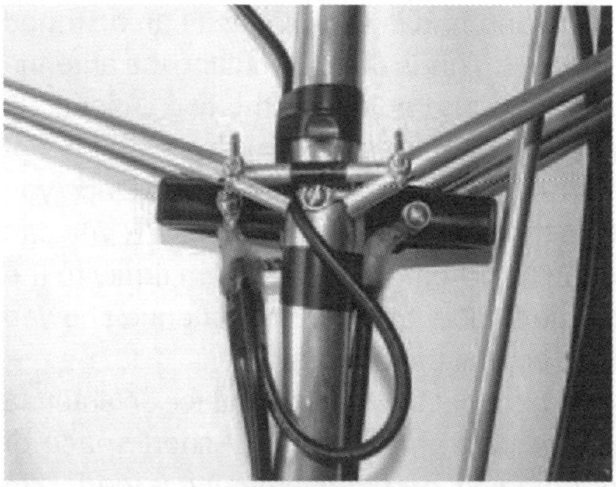

Another view of the driven element hinge. Note the insulating tape over the central bar.

Mounting clamp, made from two exhaust clamps.

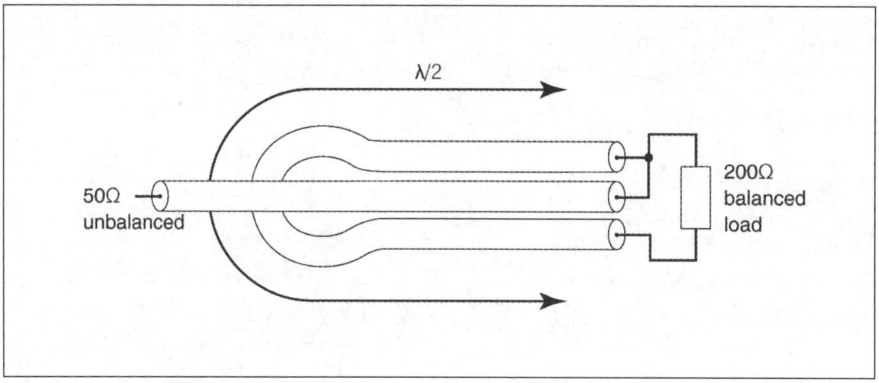

Fig 5.4: Balun details (see text).

diameter of the bar was chosen so that it would slide inside the original tube, like a trombone, so that the length could be fine-tuned for lowest SWR. **Fig 5.3** shows the final dimensions and drilling details. It's important not to drill the U shaped piece until the antenna is finished and has been adjusted for lowest SWR. The photo above left shows a close-up view of the finished assembly.

The balun was a coaxial 4:1 type, see **Fig 5.4**. The loop should be an exact electrical half-wave length; remember to allow for the velocity factor of the cable. When fixing the balun and feed line on the boom, run it symmetrically under the boom and to the rear. Terminal tags were used to connect the cables to the wing nut bolts of the driven element. The coaxial ends were made waterproof with silicone.

A mounting clamp was made from two exhaust clamps with a short piece of tubing between them (see the photo above right) and is flexible enough to allow the antenna to be used with horizontal or vertical polarisation. It can also be tilted up or down according to requirements.

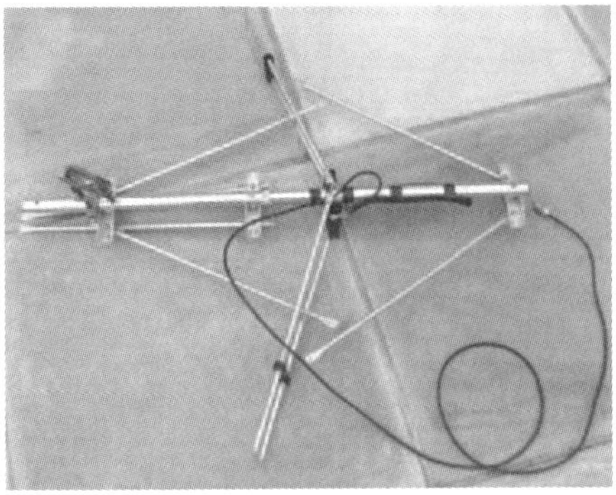

Folding the antenna.

Once the antenna is completed it can be tuned for lowest SWR. This is done by putting the antenna on a short mast and adjusting the two U sections for lowest SWR. Make sure that the two sides of the driven element are the same length. Once you have found the positions for lowest VSWR you can drill through the U bars and bolt them firmly to the driven element tube. 2W0WXM's antenna showed a 1:1 SWR across the 2m band.

Eric said his objective was to produce an antenna that could be folded up in a very short space of time and be easily and conveniently carried. "I'm pleased to have a beam antenna that can be folded into a size that easily fits into a single tent bag. It only takes a couple of minutes to open up and it's ready for use," he wrote.

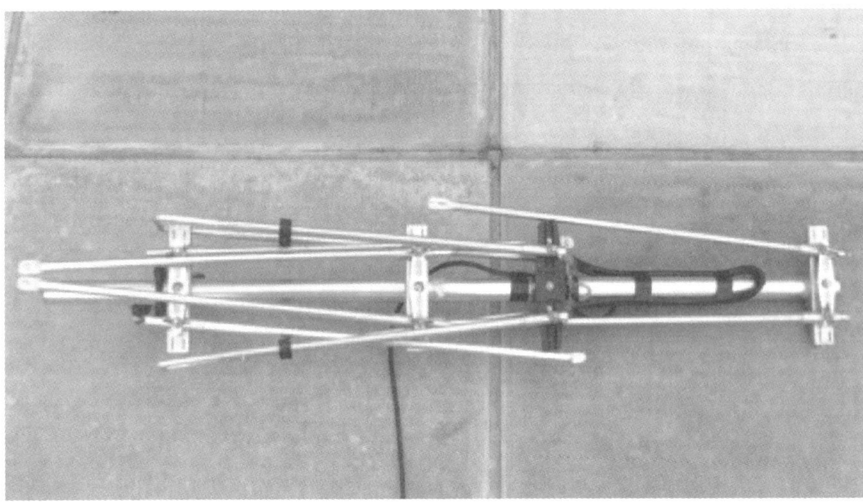

The antenna completely folded, ready for transport to the portable location.

A PARASITIC LINDENBLAD ANTENNA FOR 70CM

Anthony Monteiro, AA2TX
QST, February 2010

A Lindenblad antenna is circularly polarised and has an omnidirectional radiation pattern. It is especially useful for portable satellite operation because it eliminates the need for an azimuth or elevation rotator system. A traditional Lindenblad antenna uses four dipole driven elements to create the radiation pattern but its construction is challenging in order to get a good match to 50Ω. The 70cm parasitic Lindenblad described here provides a simpler way to match a Lindenblad to 50Ω.

The parasitic Lindenblad antenna uses a single dipole driven element with a circular 'polariser' made from parasitic elements. Parasitic elements in Yagi beams are always in the same plane as the driven element because their purpose is to improve the gain or front-to-back ratio of the antenna. In the parasitic Lindenblad, however, all of the antenna elements are in different planes.

The Lindenblad's driven element is a vertical dipole, while the parasitic elements are set at 30° from the horizontal.

The driven element is a vertical dipole extending up from the mast, as shown in the photo. The parasitic elements surrounding the driven element convert the vertically polarised signal to circular polarisation.

The parasitic elements are arranged in a similar manner to the driven dipoles of a traditional Lindenblad antenna. They are positioned equally around the centre dipole and canted at 30° from the horizontal. Because of this 30° shift, the parasitic elements absorb power from the electromagnetic field of the driven element and this causes a current to flow in them. The induced current flow causes the parasitic elements to radiate their own electromagnetic field just as if they had been driven from a feed line. Since they are arranged as a traditional Lindenblad antenna, the resulting radiation is circularly polarised.

The overall antenna pattern is the combination of the patterns of the parasitic elements and the vertical dipole. The spacing of the parasitic elements from the driven element is designed so that they absorb and re-radiate half of the total antenna power. The tuning of the parasitic elements is designed so that their vertical component cancels half of the driven element field. The resulting combination is only circularly polarised and opposite sense from the parasitic elements alone. This means that with the parasitic elements tilted as in a left-hand polarised Lindenblad, the antenna pattern will be right-hand circularly polarised.

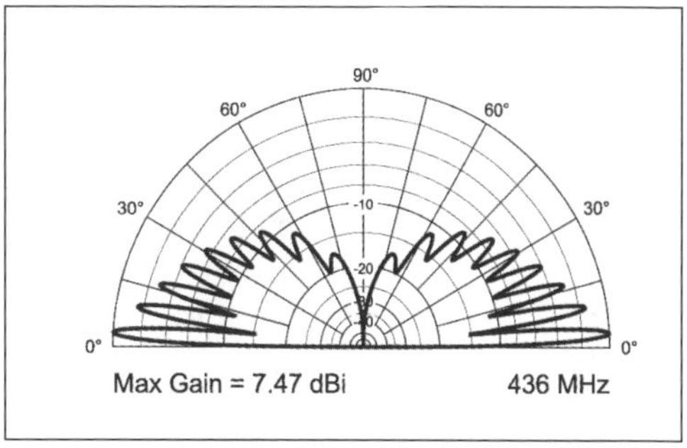

Fig 5.5: The EZNEC predicted elevation pattern of the parasitic Lindenblad. The azimuth pattern is almost perfectly circular.

The antenna elevation pattern predicted by an *EZNEC* model is shown in **Fig 5.5**. This is the right-hand circularly polarised gain with the antenna mounted 10 feet above ground. The maximum gain predicted by the model is 7.47dBic (decibels compared to an isotropic circularly polarised antenna) although the real antenna gain will be more like 6dBic. The predicted azimuth pattern, not shown, is almost

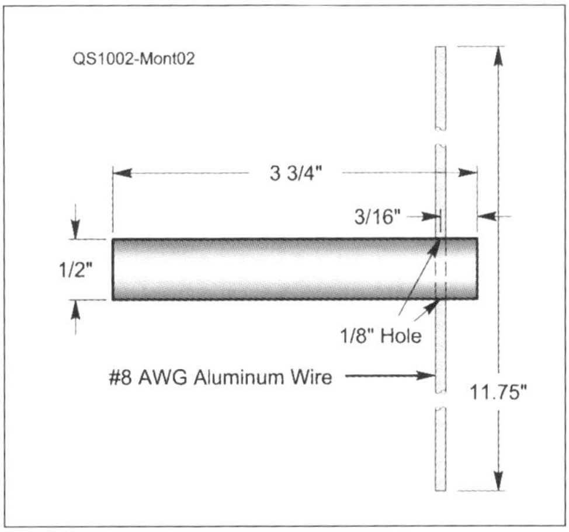

Fig 5.6: Fabrication details of parasitic element.

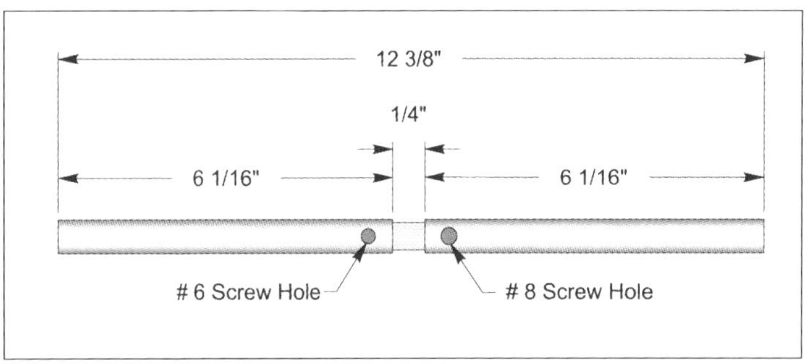

Fig 5.7: Details of the driven element construction.

perfectly omnidirectional with less than 0.1dB variation.

The driven element is designed to provide 50Ω directly but it requires a balanced feed for proper operation. If the coaxial cable is connected directly to the dipole, a significant antenna current will flow over the outside of the coax shield, which will distort the radiation pattern and cause an impedance mismatch. Fortunately, it is very easy to eliminate this shield current by using a pair of inexpensive ferrite cable sleeves.

The parasitic Lindenblad made by Anthony Monteiro, AA2TX, was constructed using only hand tools. Most of the construction and materials are not critical and the method of construction can be determined from the diagrams and photographs.

The structural elements of the antenna are made of grey-coloured UV-resistant PVC components that are cemented together. They were all found at a local hardware store. LMR-240 coaxial cable was used because of its low loss, low cost and wide availability but any small diameter 50Ω cable may be substituted as long as it is kept short.

The polariser structure supports the four parasitic elements which AA2TX made made from #8 AWG aluminium ground wire. See **Fig 5.6**.

Next he made the driven dipole assembly as shown in **Fig 5.7**. The coax is connected as shown in the photo. The cable ferrites are slipped over the open end of the coaxial cable and aligned with the bottom of the driven dipole. Fasten the ferrites and cable with wire ties.

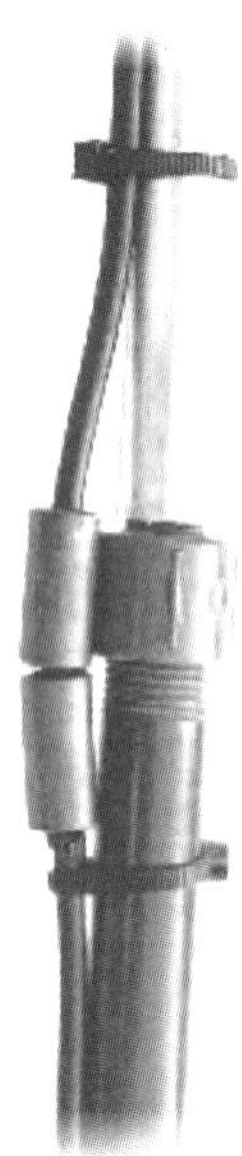

Far left: View of driven element coax cable attachment.
Left: The ferrite bead common mode choke at the base of the driven element.

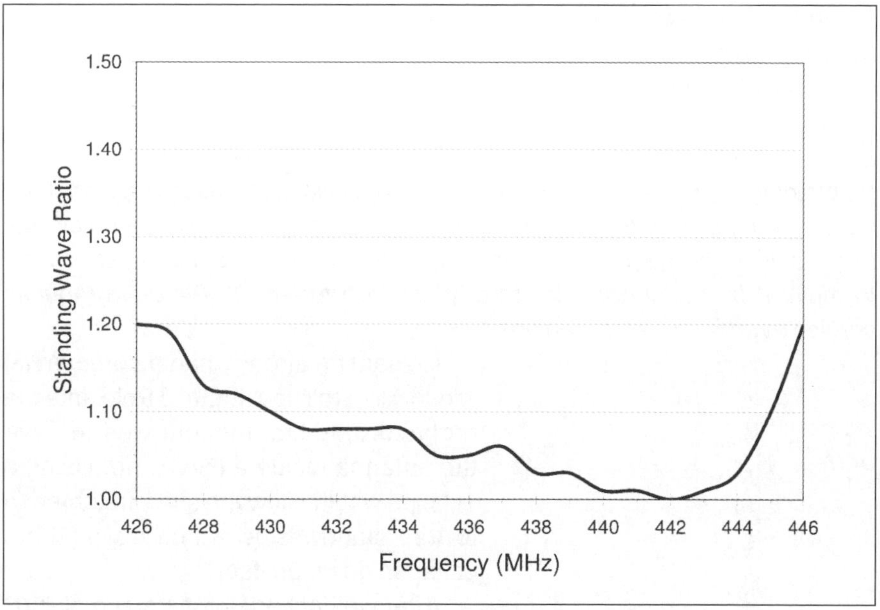

Fig 5.8: Measured SWR versus frequency of the completed parasitic Lindenblad antenna.

The parasitic element hub needs to fit over the driven element dipole so that it is centred on the PVC insulator.

Attach a PVC tube to the end of the dipole: the PVC 'riser' allows the antenna to be attached to a mast with a pair of U bolts.

The impedance match to 50Ω was checked using an antenna analyser. The antenna was connected to the analyser with a 10-foot length of coax cable. The SWR was measured over a 20MHz range centred at 436MHz with the results shown in **Fig 5.8**.

As can be seen in the chart, the antenna provides an excellent match over the satellite sub-band from 435 to 437MHz and the SWR is very low over the entire measured 426 to 446MHz range. The low SWR allows this antenna to be used as a general-purpose antenna for 70cm including CW, SSB, FM and repeater use.

The parasitic Lindenblad antenna was designed to handle up to 75W as is typical of 70cm base transmitters. This antenna should not be used with a high-power amplifier unless a different choke is employed, as the ferrites might get hot and damage the coax.

The parasitic Lindenblad antenna has been used to make SSB and FM voice contacts via the AO-7, FO-29, SO-50, AO-51 and VO-52 satellites during Field Day operations. Satellite uplinks were run at 50W output. On the downlinks, a low noise (1dB NF) pre-amp mounted at the antenna was employed. The parasitic Lindenblad has performed well during ARRL Field Day, which is an excellent test of an antenna as it is probably the busiest time of the year on the satellites.

A FREESTANDING 2-METRE J POLE

David Hannon, KE7TTT
QST, March 2009

David Hannon, KE7TTT, was a newly-licensed Technician class amateur (approximately equivalent to the UK Foundation licence class) in Washington state when he wrote this article. He said, "I wanted to figure out a quick way to get a window line J Pole antenna up in the air on a mast and yet be able to move it by passenger car to any location for ARRL Field Day, public service events and emergencies."

He said he already had a commercial window line roll up 2-metre J Pole antenna on hand. "My first thought was to have the antenna inside a PVC pipe. Using a 1in pipe would allow the stiff wire antenna to free stand inside, since it would not collapse down on itself."

After we hacked up one piece of PVC trying to fit the SO-239 coax connector inside it, we realised that there was not room to do what we wanted inside a 1in piece of pipe.

I next built on an idea from John Moore, K7NIA, who had taped one of these antennas on to the top half of a fishing rod blank. I followed suit, taping the J Pole to the outside of the upper section of 1in PVC pipe. This should allow the antenna to work as it was originally designed, with no change in SWR, as there might have been with mounting it inside the PVC pipe.

We used heavier 1.25 inch schedule 40 PVC pipe for the four-way base sleeve and legs. Each leg is about 3 feet long. It is sturdy and will sway quite a bit without tipping over, even in moderate breezes. For heavier conditions, weights or sandbags can be placed on the legs for additional stability.

We considered various ways to fabricate a five-way centre support to hold our base legs and vertical 1 inch PVC pipe all together. We took a 1.25in four-way slip sleeve and drilled a hole dead centre through it. We also drilled the same

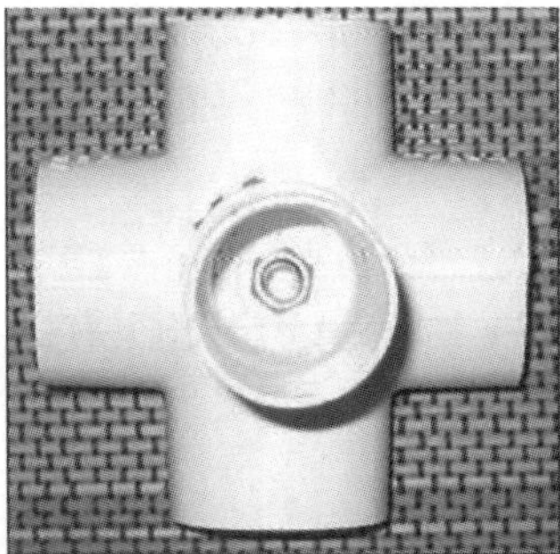

Left: The bolt through the end cap and four-way fitting.

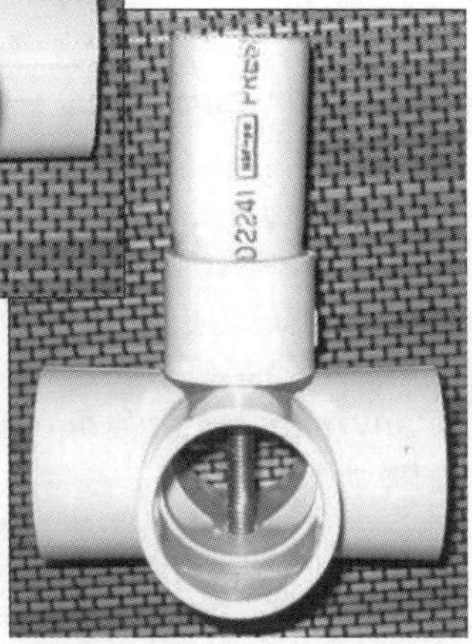

Right: The five-way hub with added stub of PVC pipe.

Above: The completed base assembly ready for the portable site.

diameter hole in the center of a 1 inch end cap. We used some epoxy putty and a stainless steel carriage bolt, flat washer and nut to hold it together as shown in the photo top left.

We didn't think the end cap had enough depth to hold up approximately 10 feet of 1 inch PVC securely, so we cut the bell end off of a 1 inch length of PVC pipe and glued the small diameter end down into the 1 inch end cap (see photo above, centre). We now have more depth for the vertical PVC to slip over.

When you do this, be sure to put it straight in. Ours was a few degrees off, causing a little lean at the top of the vertical. The completed base is shown in the photo top right.

All pieces except the small collar and five-way base assembly are slip fitted together so it comes apart for transportation and storage as shown in the photo to the right. It all fits into a large yard umbrella carry bag.

Standing back and looking at it all put together I thought it was not bad for my first attempt at building something.

I think my next freestanding 2-metre J Pole will be made from 2 inch PVC pipe. I will then be able to mount everything inside and see

All the pieces come apart for easy transportation and storage.

how that works. The 2 inch diameter should provide enough room to mount the SO-239 through the wall of the PVC pipe. I think it would be appealing to just have a PVC stick up in the air as an antenna.

PORTABLE VHF/UHF YAGI FOR SATELLITE WORKING

Thomas M Hart, AD1B
QST, May 2008

The antenna described here was designed for working through satellites while out portable using a hand-held transceiver. However, the basic concept can be used for 'fox hunting' or any other portable operation when additional gain is required from a hand-held transceiver.

Thomas, AD1B, wrote that he had been operating through the SO-50 and AO-51 satellites using a dual-band Yaesu FT-60 handheld transceiver. The standard flexible antenna worked, but to improve conditions he built a dual band 2m and 70cm Yagi without driven elements, matching network or feed line. He mounted the FT-60 on the antenna boom, using the handheld itself as the driven element.

"After testing several configurations on Roy Lewallen's

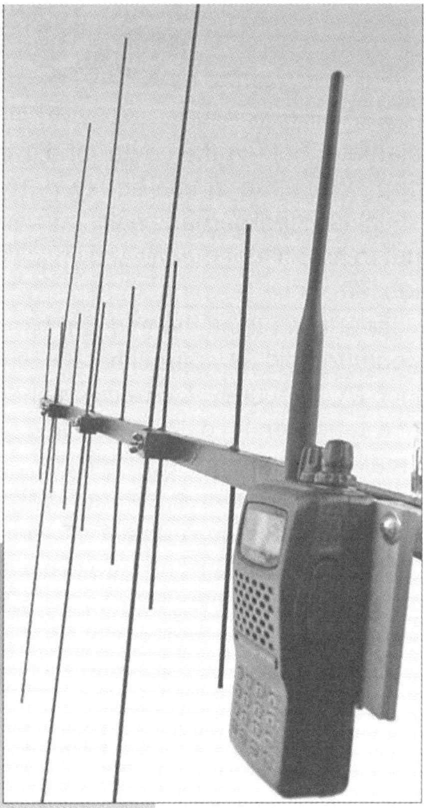

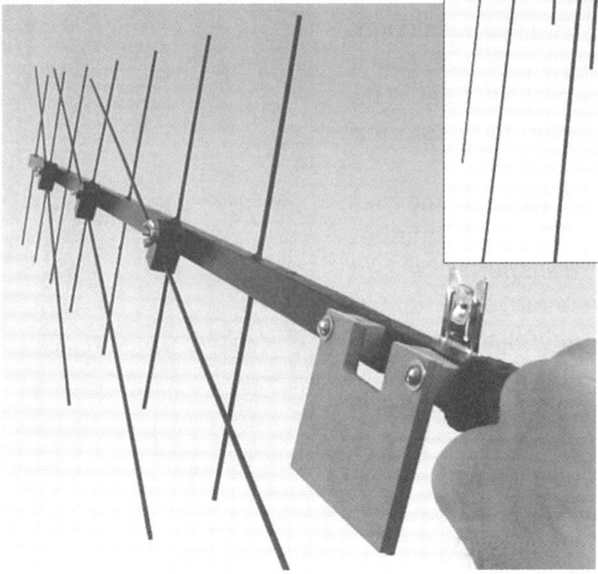

Above: The completed antenna, with dual-band handheld in place.

Left: Dual band 'Handy Yagi' with 2m elements folded for storage.

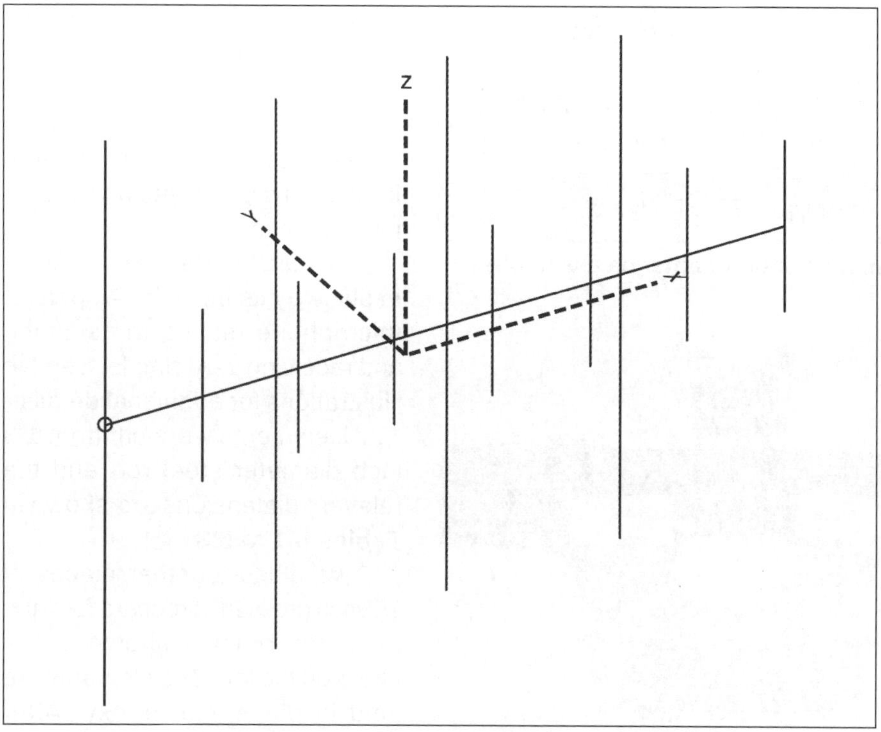

Fig 5.9: An *EZNEC*-generated diagram of the antenna shows the element placement. The Yaesu FT-60 is represented by the half-wave loaded element on the left.

EZNEC program **[1]**, I settled on a design with seven directors on 70cm and three on 2m", he wrote (see **Fig 5.9**).

The final design balances performance and size. The elements for both bands are interlaced and mounted in parallel. There is no driven element or reflector. Instead, the FT-60 and a bicycle handlebar grip occupy the usual reflector and driven element end of the boom. The 2m elements can be rotated parallel to the boom to simplify storage.

Director:	D1	D2	D3
Length:	37.5	36.375	36.0
Elements:	DE to D1	D1 to D2	D2 to D3
Spacing:	12	12	12
Cumulative:	12	24	36

Table 5.1: Length and spacing of 2m elements (inches).

Director:	D1	D2	D3	D4	D5	D6	D7
Length:	11.750	11.688	11.625	11.563	11.500	11.438	11.375
Elements:	DE to D1	D1 to D2	D2 to D3	D3 to D4	D4 to D5	D5 to D6	D6 to D7
Spacing	6.78	6.78	6.78	6.78	6.78	6.78	6.78
Cumulative:	6.78	13.56	20.34	27.12	33.9	40.67	47.45

Table 5.2: Length and spacing of 70cm elements (inches).

Boom	**0.75 x 0.75 x 55**
Handle grip	**5**
Handheld bracket	**3.5 x 3.0 x 0.25 (WxHxD)**
Notch for handheld	**1.125W x 0.75H**
Bracket spacers	**0.375**
2m element mounts	**0.75 x 0.75 x 1.5**

Table 5.3: Dimensions of other antenna assemblies (inches).

The mounting bracket, spacers and mic holder.

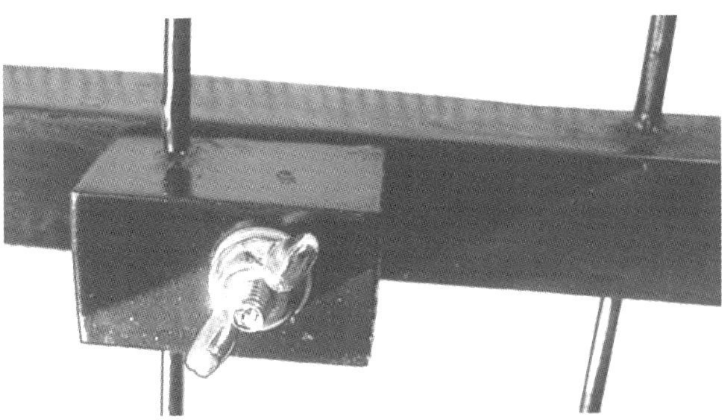

The wing nut and bolt allow the 2m element to rotate.

A screw eye at one end can be used to hang the antenna when not in use. Computer modelling indicated that λ/4 element spacing works reasonably well on both bands. This allows the use of a 55 inch boom.

The handheld is attached firmly in place by its belt clip. A speaker microphone makes transmitting and receiving very simple. See the illustrations for additional details.

All elements were cut from 1/8 inch diameter steel rod and the relevant dimensions are shown in **Tables 5.1** to **5.3**.

Two glued-together pieces of 1/8 inch pressed fibreboard form the handheld bracket shown in the photo on the left. The elements are held in place with epoxy. After cutting all 10 to length, drill seven holes through the boom and three more through the 2m element mounts. Slide the elements into place and apply a bead of epoxy on both sides. Several light applications of epoxy will hold the rods in place. All 2m element mounts require two holes. One secures the element while the second is used for the bolt and wing nut that allow the element to rotate during storage or transport, as shown in the photo to the left.

Finally, two coats of black satin paint give the antenna a more professional appearance.

Faced with an absence of analytical instruments, the testing process involved *EZNEC* computations and operational observations. Field testing on 2m involved contacting the 146.97MHz repeater in Paxton, Massachusetts. The repeater, according to my Magellan Topo 3D GPS software, is 38 miles away at a bearing of 251°. The FT-60 helical antenna cannot reach the repeater from my house. With this dual-band handy Yagi, I had no trouble reaching the repeater. Moving the antenna in an arc towards

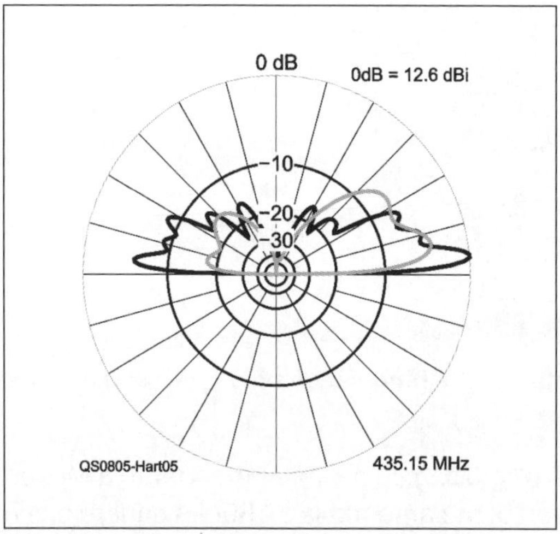

Fig 5.10: Predicted elevation patterns of the 2m (grey) and 70cm (black) Yagis.

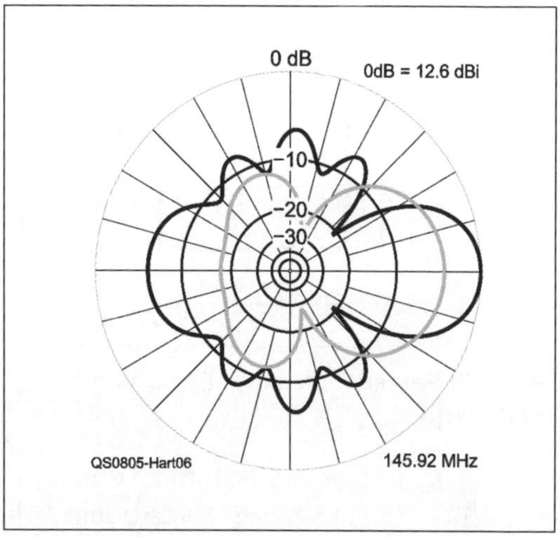

Fig 5.11: Predicted azimuth patterns at the peak of the elevation response for the 2m (grey) and 70cm (black) Yagis.

and away from the repeater produced corresponding signal strength changes on the meter corresponding to the predicted directivity.

On 70cm, field tests involved reception of signals from the AO-51 and SO-50 satellites. Signal strength increased dramatically when the antenna approached and centred on the target.

EZNEC computed the front-to-back ratios as 14dB for 2m and 5.5dB for 70cm. *EZNEC* azimuth and elevation patterns are shown in **Fig 5.10** and **5.11**. My conclusions are that operations on both bands are in agreement with *EZNEC* predictions.

COLLAPSIBLE 6M 2-ELE BEAM FOR PORTABLE USE

Mike Parkin, G0JMI

RadCom, September 2012

Mike, G0JMI, developed this collapsible 50MHz two-element wire beam antenna for portable use. It fits into a space of approximately 25cm x 25cm x 1m long, making it suitable to carry in the boot of a small car.

Mike wrote, "The original intention was for a traditional two-element 50.15MHz beam with a spacing of 1.5m used between the reflector and driven element (about 0.25 wavelength), as shown in **Fig 5.12**. I realised that the antenna needed less space if the elements

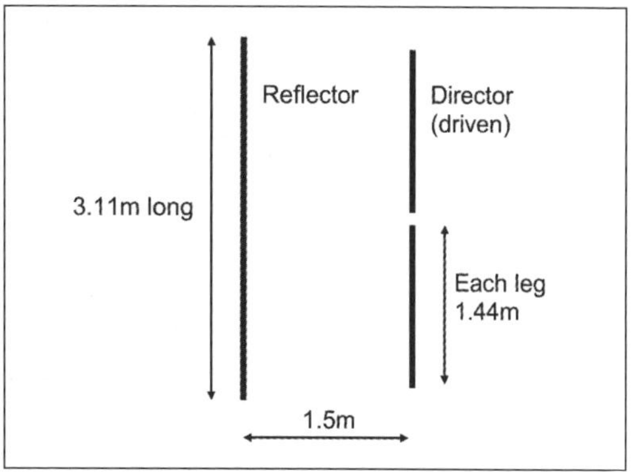

Fig 5.12: Original 2-ele Yagi concept.

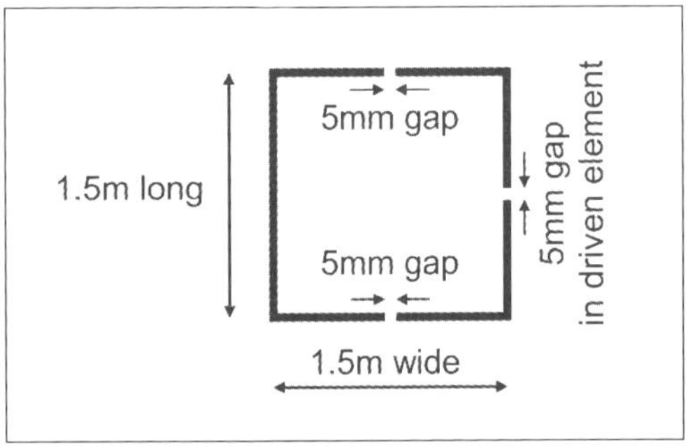

Fig 5.13: Folded 2-ele antenna concept (see text).

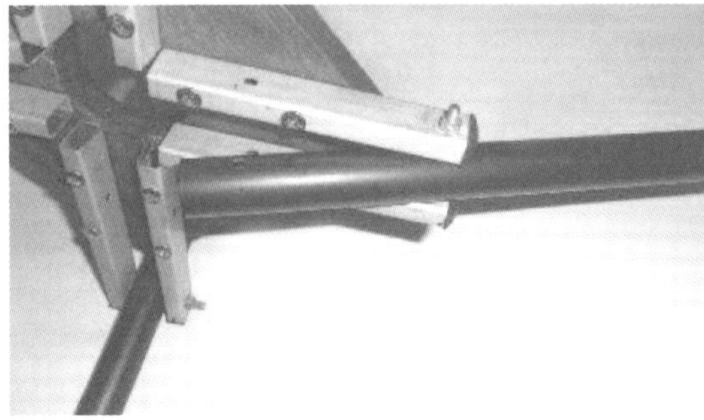

Detail of the hinge assembly (see text).

are folded into an oblong shape **(Fig 5.13)**, in a similar manner to a Moxon or VK2ABQ type beam." The heart of the antenna is a 150mm square piece of plywood, to which four 1m lengths of plastic conduit are hinged. This allows the antenna to be folded up. The antenna elements are supported by the lengths of conduit. The photo above shows the general arrangement of the homebrew hinges.

First, a 50mm hole was cut in the centre of the plywood board to allow the antenna to be attached to a mast, described in more detail later. Eight runners were made up to form the hinges for the plastic conduit legs. These were 11cm long and fabricated from 10 x 15mm scrap aluminium oblong tubing. Batten wood should work equally well if suitable aluminium oblong tubing is not available.

Four holes were drilled in each runner as follows: starting from one end, drill two 5mm holes at 15mm and 50mm right through the narrower sides. These holes will allow each runner to be screwed to the plywood board. Then drill two 4mm holes through the longer sides at 102mm and 32mm along each runner.

Each hinge was formed using two runners, screwed on the plywood board 21mm apart, aligned equally either side of the diagonal across the square board, with the ends flush with 50mm central hole. This leaves 45mm of runner protruding from the plywood board's edge, with a 4mm hole at each end. These are the hinge holes for the plastic conduit legs.

Each plastic conduit leg was drilled with two 4mm holes, at 18mm and 88mm, to match up with the two 4mm holes within each runner assembly. A suitably long nut and bolt was used to secure each plastic conduit leg, so forming the hinge at the protruding end of the runner assembly.

Holding the plywood board with the hinges containing the plastic conduit legs underneath, the conduit legs self-lock against the plywood board under gravity, forming the X frame for the antenna. If required, a securing nut and bolt can be used to fix each conduit leg in place using the second 4mm hole within each hinge assembly. When completed, the total diagonal length across the X was 2.08m. 20mm plastic conduit connector pieces were used to hold the wire beam in place at the end of each conduit pipe leg. A 4mm hole was drilled through the end of each connector piece to allow the wire to be threaded through. This allows ease of removing the antenna from the X frame to store it away when not in use.

Attaching the antenna to a pole using a hose clamp.

The upper surface of the plywood board has two mild steel metal brackets mounted 90° apart with the 'fold' of each bracket set flush with the 50mm central hole. This allows the antenna to be mounted on a mast using a suitable clamp, for example a hose clamp. The mast mounting arrangements are shown in the photo to the right.

The driven element is a half wavelength dipole cut for 50.150MHz. The length was found, from experimentation, to be 2.88m for the 5A-rated copper single stranded cable I used. The reflector was made from the same cable but cut about 8% longer than the dipole, at 3.11m in length.

Before making up the beam, the four 20mm plastic conduit connector pieces mentioned earlier were threaded on to the legs of the dipole (one each) and reflector (two used) before proceeding further.

Three 5mm wide by 25mm long insulators were made using single-sided copper-clad board with a 5mm strip of copper removed centrally across the width (to form each insulator). Two insulators are used between the reflector and the dipole. Before assembly, two 75mm lengths of suitable heatshrink tube were threaded on to the wire, one per insulator. The 5A wire was stripped back and soldered on to each insulator, as shown in the photo below left. After soldering, the heatshrink was moved over the insulator and shrunk to provide waterproofing.

The third insulator forms the centre for the dipole. The dipole insulator had two brass screw-connectors soldered on to allow the feeder to be attached (below right). These brass connectors were obtained from a cut-up 'chocolate block' connector strip. A suitable plastic box with snap-on cover was used to protect the dipole's connections. Tight-fitting holes were drilled to allow cable access.

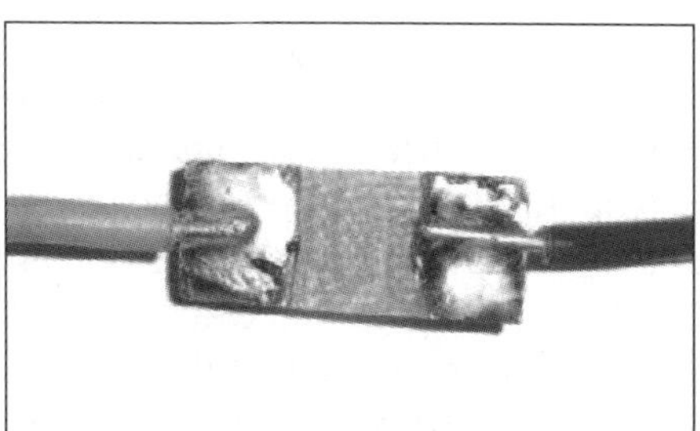

Insulating spacers made from PCB material allow an accurate 5mm gap to be maintained.

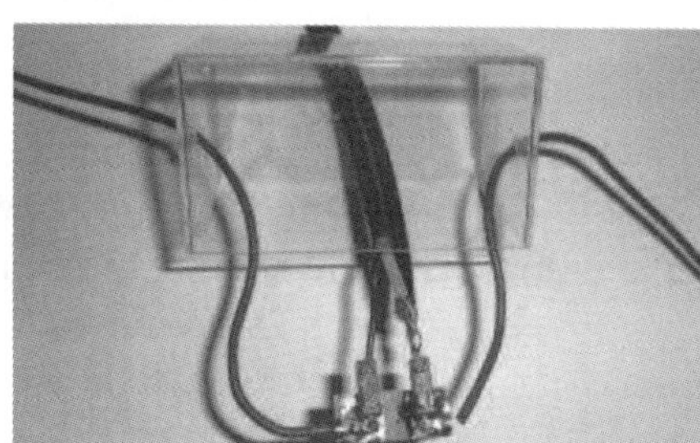

How the end of the feed and balun coax join the driven element. See also Fig 5.14.

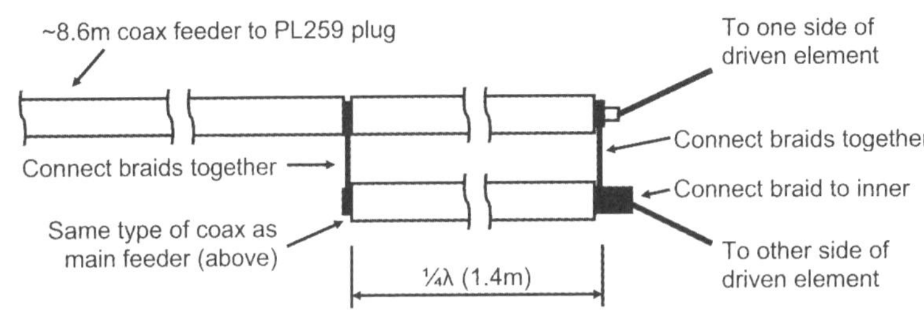

Fig 5.14: Connections between the feed coax and balun coax (not to scale).

The beam antenna was then installed on the X former by pressing the 20mm plastic conduit connector pieces on to the ends of the conduit pipes. Once measured up for position, white marks were painted on the wire inside each conduit connector piece to show where the beam antenna needed to be aligned for use. The total distance between the ends of each diagonal forming the X with the antenna attached then extended to 2.12m.

10m of RG58U 50Ω coax cable was used as the feeder for the antenna, with a PL-259 plug fitted at the radio end. A quarter wavelength balun was used to match the unbalanced feeder to the antenna's dipole. This was made up from the same RG58U cable. From experimentation, the length of the balun cable was found to be 1.4m, to allow for the cable's velocity factor at a test frequency of 50.150MHz.

Arrangements for the balun are shown in **Fig 5.14**. The balun assembly was held together using lengths of insulation tape, spaced about 12cm apart. The connection between the coax outer shield and balun cables was soldered and then waterproofed using insulating tape.

The antenna was found to give a VSWR of under 1.4:1 between 50.001 and 50.299MHz when mounted horizontally about 1.5m off the ground. Increasing the height to 4m improved the VSWR to better than 1.2:1 across the band. The tests were done at a power of 10W. To provide a safety margin for the antenna, it is suggested that a maximum power of 50W is not exceeded.

A horizontally polarised 50.210MHz signal source was set up at a distance of about 20m from the antenna, which was also horizontally polarised and mounted at 1.5m above the ground, on a tripod. With the antenna facing the signal source, the signal source power level was adjusted to read S9 on an Alinco DX-70 tuned to 50.210MHz in CW mode. The antenna was turned at 15° intervals to allow a series of S-point measurements to be taken.

These were plotted, to provide an indication of the polar diagram for the antenna. This test indicated a front-to-back difference of about 14dB. A similar test was then repeated, using a commercial 2-element Yagi. Performance is similar, although the 2-element Yagi had a slightly better

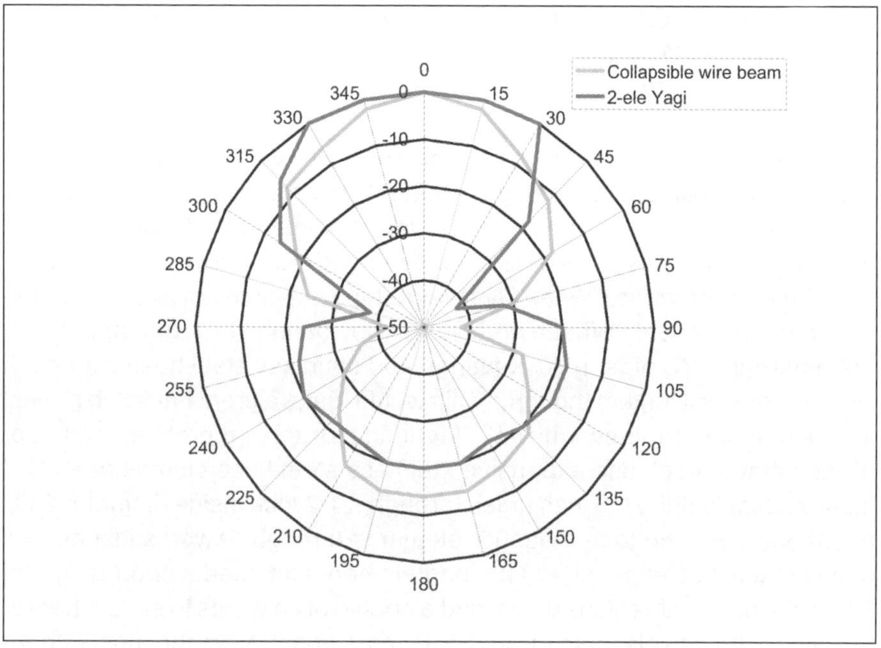

Fig 5.15: Polar patterns of the wire beam antenna compared with a commercial 2-ele Yagi. Scales are degrees and 0 to –50dB.

front-to-back ratio. The results have been normalised and are shown on the same polar diagram, **Fig 5.15**.

During the Sporadic E season, the antenna was used to make SSB QSOs from my QTH in IO91 square with a SV9 station in Crete (KM25) and a 5B4 in Cyprus (KM64) with received signal reports of 59. Further to this, a UX1 station in the Ukraine (KN29) was worked on the key with a received signal report of 559. The power output used to work these stations was 10W. Compared to other stations worked and signal reports received using a 6m dipole, the 2-element collapsible wire beam has made an appreciable difference.

The completed 6m 2-ele collapsible antenna.

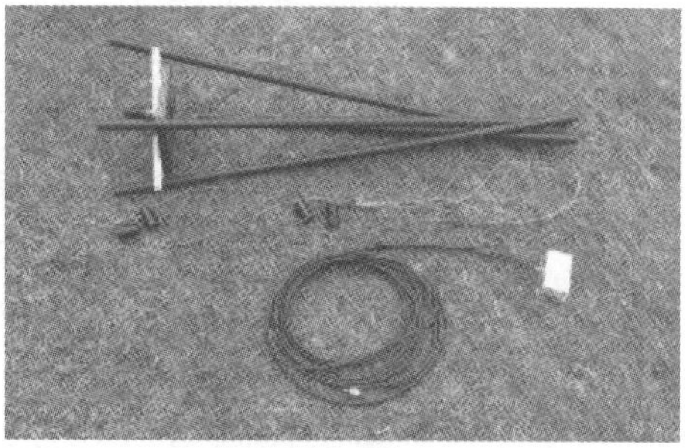

The antenna folds down to a compact size and shape for transport and storage.

A 2M AND 70CM PORTABLE TAPE MEASURE BEAM

John Portune, W6NBC

QST, January 2012

Many amateurs have made 2m tape measure beams for ARDF transmitter hunting. In this article that design is taken two steps farther. First, the cost is reduced and satellite capacity is added. A bonus is that a double boom design lets you fold the antenna up conveniently for transport or storage.

John Portune wrote, "When I first saw a tape measure 2m beam for transmitter hunting, I said: 'What a great idea'. So I bought a kit with tape measure material, PVC pipe, PVC X fittings and stainless steel hose clamps. It worked very well. Later I thought: 'This would make a great ham club project if I could just get the cost down'." Lightning struck – there's an easy approach that doesn't use expensive stainless steel hose clamps or PVC X fittings. Simply drill 7/16 inch holes through a 1/2 inch inside diameter PVC boom and push the tape measure elements through. It works just as well and is much less expensive. The long elements do need a short length of 3/8 inch wood or fibreglass dowel and a couple of tie wraps to stiffen them in the wind. The dowels push through the same holes. Now the most expensive parts of the beam is the RG-58 coax and the connector.

More recently I became interested in working Mode J OSCAR satellites – 2m FM phone towards the satellite, 70cm FM phone from the satellite. What kind of antenna should I use? At a local club meeting I learned that

The dual purpose, dual boom portable tape measure Yagi for 2m and 70cm.

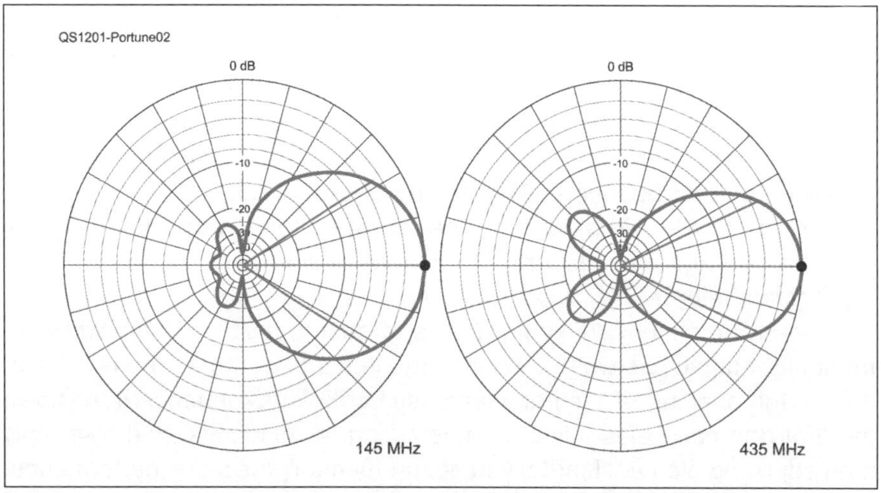

Fig 5.16: *EZNEC* elevation plots with both Yagis vertically polarised. As shown, the 145MHz gain is 7.5dBi, the front-to-back ratio is 29dB. For 435MHz, the gain is 10.4dBi, the front-to-back ratio is 38dB.

there are several commercial handheld satellite beams on the market. The lecturer gave us a live demo of one during an actual satellite pass. But were there any good homebrew designs? Then I remembered my handy little 2m tape measure transmitter hunting beam. Why not just add a 70cm beam to it?

The one shown is the result. I did not make any changes to the existing three element 2m transmitter hunting beam. For though it is a little shorter than most commercial satellite beams, and is optimised more for front-to-back ratio, it works just fine for the 'birds' and of course still for transmitter hunting. I did, however, optimise the new 70cm beam for forward gain. **Fig 5.16** shows *EZNEC* **[1]** elevation plots for both, with the beams turned vertically.

Booms rotated 90° and 2m elements folded back. Note tie wraps to hold ends.

As just a bit of simple theory, the 70cm beam needs to be rotated axially 90° from the 2m beam. This makes the two 'invisible' to each other. I've seen designs in which they are in the same plane. I tried this and found it unsatisfactory. The reason, I believe, is that a 2m Yagi has a third harmonic resonance near 70cm. The directivity of the 2m Yagi on UHF is poor. If you mount the two beams in the same plane, the patterns clash. I modelled this with *EZNEC* and saw that it is a poor idea.

A particularly handy feature of this design

is the dual boom. I had at first thought to just drill extra holes in the existing single boom for the 70cm Yagi. In terms of performance, that would have been fine. But on the 2m only version I had been folding the element ends back under tie wraps for transport or storage. Had I added a 90° fixed position UHF beam, the whole structure would have become cumbersome.

Two rotatable booms is a better approach. By holding the booms together with two tie wraps one can easily rotate the beams into the same plane for storage or transport. The folded-up dual-band beam is now no bigger than before.

One of the slightly tricky parts was to design a 70cm beam with elements that fit well between the existing elements of the 2m beam. Fortunately, the spacing of Yagi elements isn't critical. Experience has shown me that one can select almost any spacing, within limits, and by simply then adjusting element length, you get pretty much the same performance. **Fig 5.17** gives placements and sizes. The elements are made from ordinary small-width tape measure material.

The impedance at the centre of the 2m beam will be less than 50Ω. By shortening the driven element and inserting an inductive hairpin in shunt with the feed point, the impedance will be raised to 50Ω resistive. Make an inductive hairpin from #12 AWG solid copper wire, 8.5 inches (22cm) in length. It yields reasonable SWR and is easily adjusted. The 70cm Yagi's driven element naturally matches 50Ω without any added tuning elements.

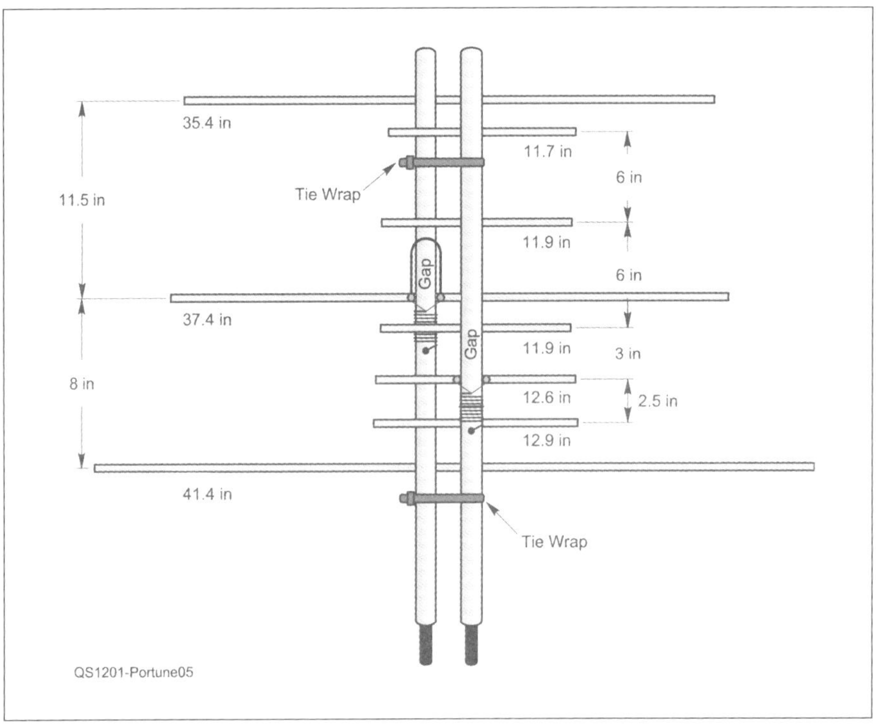

Fig 5.17: Dimensions with booms rotated. Note the boom rotation tie wraps.

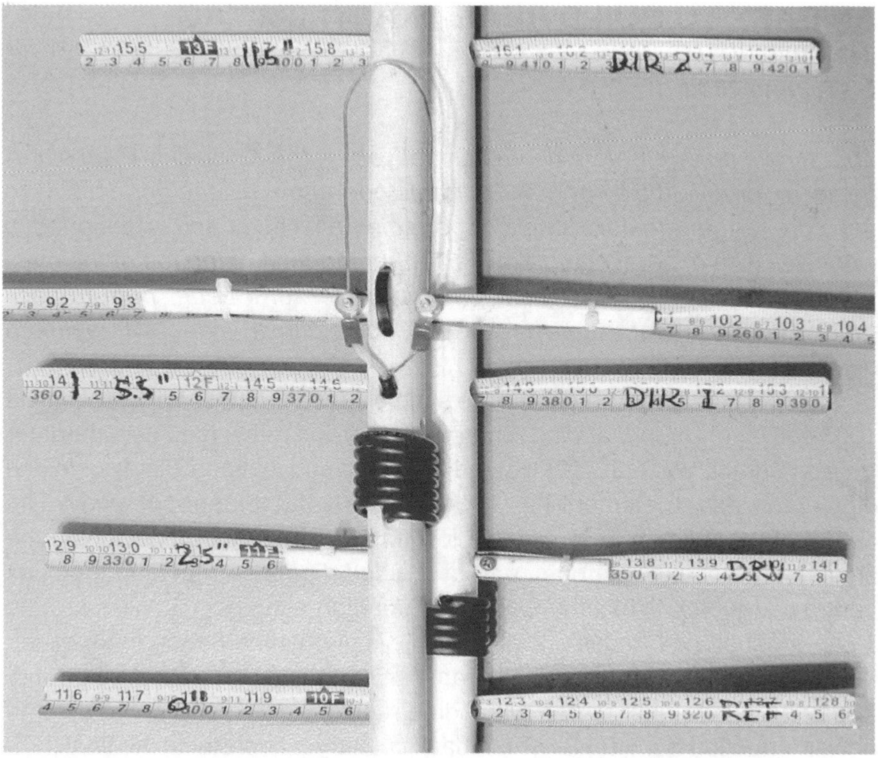

Feed point detail showing choke baluns, element stiffeners and hairpin match on the 2m beam. The booms have been rotated into 'storage mode'.

As with most coax fed antennas, baluns on both beams is highly recommended. Shown are choke baluns made by wrapping the RG-58 feed coax around the boom several times. Secure the ends through holes in the boom. For 2m, six turns is adequate; for 70cm, use four turns. Run the coax down the inside of the boom and out the end. Use two 6-32 screws and four nuts to connect to each driven element. Leave a 1/4 inch gap in the middle. Scrape the paint off the tape measure material and tin the area with solder to provide a good connection. Separate out a short length of the coax's inner and outer conductors and crimp on ring terminals.

While I normally make my OSCAR contacts using two separate handheld transceivers, one for each band, you may, of course, use a dual-band handheld, but in that case you will need a diplexer to connect the two antenna feed lines to the radio's antenna port. There are also several satisfactory homebrew diplexer designs on the Internet.

I claim no special magic for this antenna design. The Yagis are classical designs and the tape measure method well known. The charm is the mechanical arrangement. By eliminating PVC X fittings and stainless hose clamps, it is inexpensive, and by also using two booms it folds up neatly. I now can seamlessly combine two of my favourite ham radio disciplines, transmitter hunting and ham satellites in one handy portable antenna.

AN EASY TO BUILD REAR-MOUNT 2M YAGI

Zack Lau, W1VT

QST, June 2012

This simple three element beam can get you on the air quickly in an emergency or for a field Day or other portable operation. It's the perfect way to get more range with a handheld or portable transceiver and is designed to work well with a metal mast. You can either mount it at the top of the mast for maximum range, or 20 inches (51cm) lower for a better front-to-back ratio. The free space gain is about 7.5dBi with a wide 114° azimuth beamwidth for easy aiming.

Zack, W1VT, wrote: "I copied the simple J feed that first appeared in the 1963 edition of *Understanding Amateur Radio*, and reinvented decades later by Kent Britain, WA5VJB" **[2]**.Instead of using a conventional dipole or folded dipole, the driven element has half of the driven element cut away. The centre of the full dipole provides a low impedance point for attaching the shield of the coax – the open end of the half dipole connects to the coax centre conductor. While the SWR is a little high – the best SWR is around 1.5:1, most of the band is under 2:1. This is plenty good enough for excellent performance with a handheld transceiver and a short transmission line. It may be a little high to use with some mobile rigs – you may see some power reduction from the SWR foldback circuitry that protects the final transistors.

The design was a three step process. The first step was modelling the antenna with *Yagi for Windows* (one of *The ARRL Antenna Book* compan-

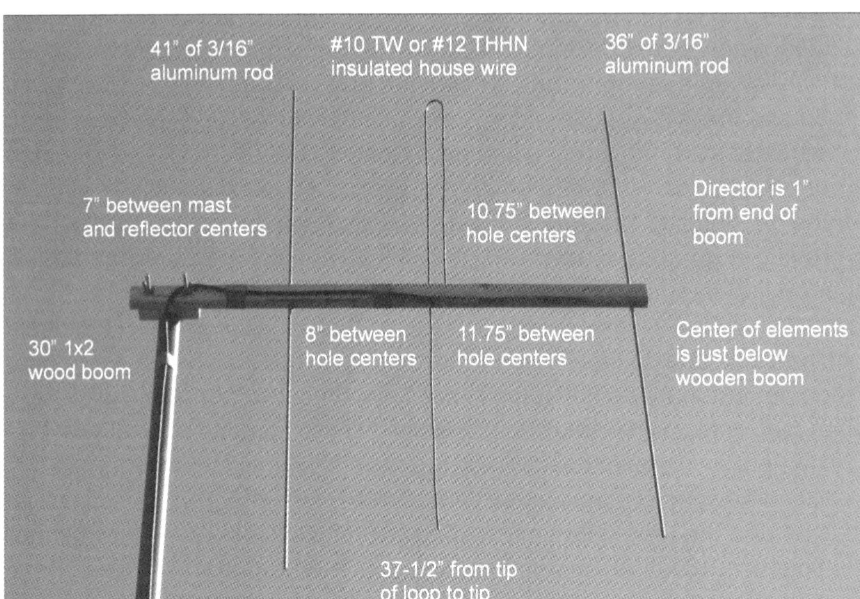

The completed antenna showing key dimensions of the 2m Yagi mounted at the top of a metal mast.

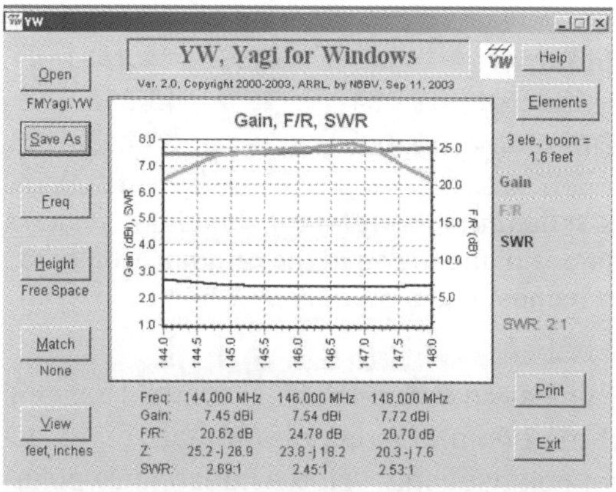

Fig 5.18: Free space gain, front / rear and SWR, according to *Yagi for Windows*.

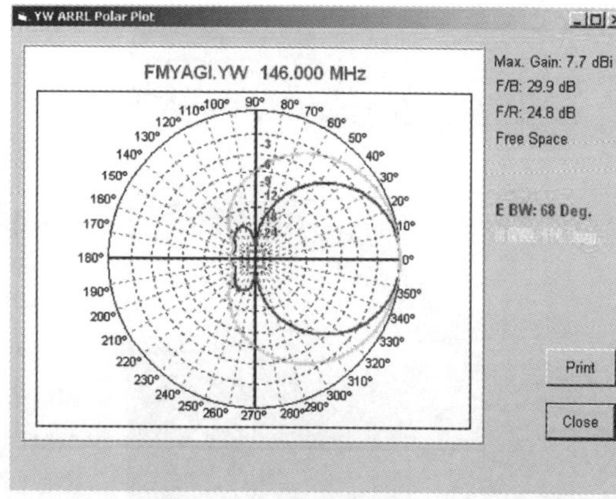

Fig 5.19: *Yagi for Windows* program output. (The azimuth pattern is grey and the elevation pattern is black. Both patterns are in free space without the mast.)

ion programs, to optimise the free space performance of the three elements over the FM part of the band (see **Figs 5.18** and **5.19**). Next, I modelled the antenna with *EZNEC Pro* **[1]** to optimise the pattern over real ground with the antenna mounted on a metal mast (see **Figs 5.20** and **5.21**). Finally, I used trial and error with an antenna analyser to design the dimensions of the J. The last step was the most time consuming of the three.

I found that the front-to-back was optimised by placing the mast 7 inches (18cm) behind the reflector, and placing the mast an inch higher than the reflector. This sacrifices 0.5dB of gain, compared to mounting the antenna for maximum height. For maximum range, you want the antenna as high as possible. But, if you have an interference issue, you can hide the antenna behind a metal mast and null out the interference.

This antenna is small and light enough that the lack of balance isn't a big issue. The rear mounting avoids the boom interaction that often

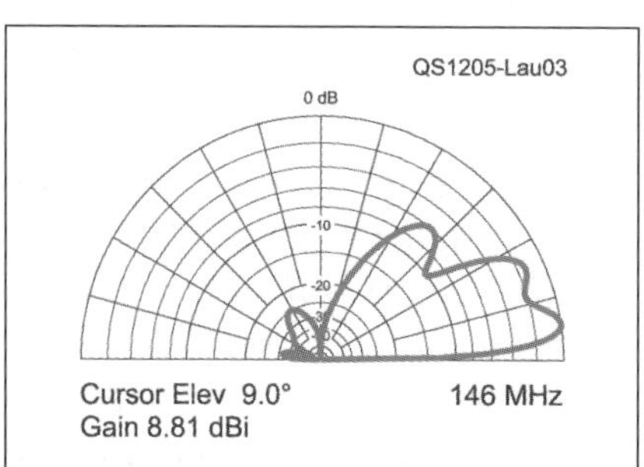

Fig 5.20: *EZNEC* elevation pattern of the Yagi on the mast over real ground. The mast reflector costs about 0.5dB of gain.

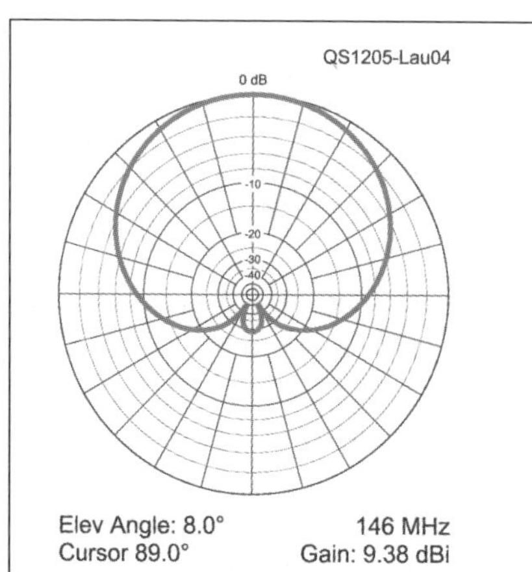

Fig 5.21: *EZNEC* azimuth pattern of the Yagi mounted on the top of a 10 foot (3m) mast for maximum gain.

Close-up of the connections to a driven element made out of #10 AWG wire.

Element	1/8	3/16	1/4
Reflector	41.034	41.0	40.956
Director	36.320	36.0	35.746

Table 5.3: Lengths for different element diameters (dimensions in inches as in the original article; to convert inches to cm multiply by 2.54.)

plagues vertically polarised beams mounted near the centre of the boom. It is important that you select the proper polarisation – using a horizontally polarised beam for FM work can cost you several S-units of signal, more than wiping out the benefit of the beam.

Merely cutting the provided antenna dimensions accurately should be enough for good performance with a handheld transceiver – the loss in range due to the SWR mismatch is negligible. But, if you wish to tweak the antenna for best SWR, I'd suggest cutting the driven element a quarter inch long and clipping off 1/16 or 1/8 of an inch at a time.

Because the width of the J is not very critical, you could narrow the driven element to compensate for cutting off too much.

The dimensions are shown in **Table 5.3** and in the lead photo. If you want to make the driven element out of bare wire, you can use #6 AWG copper ground wire with an end-to-end length of 38.25 inches (approx 99cm). A 5 foot (1.52m) length of wire will be required – one third of it is used in forming the J. The photo above left shows the details of the driven element connections to an element made out of #10 AWG wire.

A PORTABLE VHF / UHF ROLL-UP J-POLE ANTENNA

Edison Fong, WB6IQN
QST, March 2007

Ed Fong, WB6IQN, designed a dual-band J-pole antenna for base station use which he called the 'DBJ-1'. He described it in the February 2003 QST *and following publication of that article had over 500 enquiries about the antenna. The most common request from users was for a portable 'roll-up' version of the antenna for backpacking or other portable use. WB6IQN wrote: "To address this request, I will describe how the principles of the DBJ-1 can be extended to a portable roll-up antenna. Since it is the second version of this antenna, I call it the DBJ-2."*

The DBJ-1 is based on the original J-pole design. Unlike the popular ground plane antenna, it doesn't need any ground radials.

The original 2m J-pole, made of ribbon feeder, is shown in **Fig 5.22**

The dual-band version, DBJ-1 is easy to construct using inexpensive materials from your local hardware store. For its simplicity and small size, the DBJ-1 offers excellent performance and consistently outperforms a ground plane antenna. Its radiation pattern is close to that of an ideal vertical dipole because it is end-fed, with virtually no distortion of the radiation pattern due to the feed line. A vertically polarised, centre-fed dipole will always have some distortion of its pattern because the feed line comes out at its centre, even when a balun is used. A vertically polarised, centre-fed antenna is also physically more difficult to construct because of that feed line coming out horizontally from the centre.

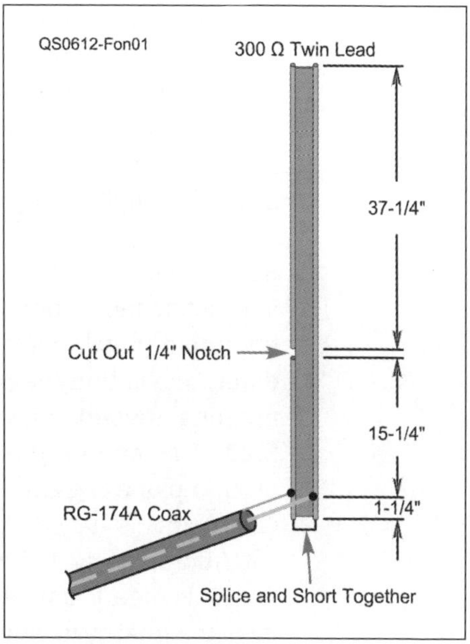

Fig 5.22: The original 2m ribbon J-pole.

The basic J-pole antenna is a half-wave vertical configuration. Unlike a vertical dipole, which because of its centre feed is usually mounted alongside a tower or some kind of metal supporting structure, the radiation pattern of an end-fed J-pole mounted at the top of a tower is not distorted.

The J-pole works by matching a low impedance (50Ω) feed line to the high impedance at the end of a λ/2 vertical dipole.

This is accomplished with a λ/4 matching stub shorted at one end and open at the other. The impedance repeats every λ/2, or every 360° around the Smith Chart. Between the shorted end and the high impedance end of the λ/4 shorted stub, there is a point that is close to 50Ω and this is where the 50Ω coax is connected.

By experimenting, this point is found to be about 1.25 inches from the shorted end on 2m. This makes intuitive sense since 50Ω is closer to a short than to an open circuit.

Although the Smith Chart shows that this point is slightly inductive, it is still an excellent match to 50Ω coax. At resonance the SWR is below 1.2:1. **Fig 5.22** shows the dimensions for a 2m J-pole. The 15.25-inch λ/4 section serves as the quarter wave matching transformer.

A commonly asked question is, “Why 15.25 inches?” Isn’t a λ/4 at 2m about 18.5 inches? Yes, but twinlead has a reduced velocity factor (about 0.8) compared to air and must thus be shortened by about 20%.

A conventional J-pole configuration works well because there is decoupling of the feed line from the λ/2 radiator element since the feed line is in line with the radiating λ/2 element. Thus, pattern distortion is minimised. But this only describes a single band VHF J-pole. How do we make this into a dual-band J-pole? To incorporate UHF coverage into a VHF J-pole requires some explanation. First, a 2m antenna *does* resonate at 70cm. The key word here is *resonate* though: any LC circuit can be resonant, but that does not imply that it works well as an antenna. Resonating is one thing; working well as an antenna is another. You should understand that a 2m λ/4 matching stub works as a 3λ/4 matching stub on 70cm, except for the small amount of extra transmission line losses of the extra λ/2 at UHF. The UHF signal is simply taking one more revolution around the Smith Chart.

The uniqueness of the DBJ-1 concept is that it not only resonates on both bands but also actually performs as a λ/2 radiator on both bands. An interesting fact to note is that almost all antennas will resonate at their third harmonic (it will resonate on any odd harmonic, 3rd, 5th, 7th, etc). The difference is that the performance at the third harmonic is poor when the antenna is used in a vertical configuration, as in the J-pole shown in **Fig 5.22**. This can be best explained by a 19-inch 2m vertical over an ideal ground plane. At 2m, it is a λ/4 length vertical (approximately 18 inches). On 70cm it is a 3λ/4 vertical.

Unfortunately, the additional λ/2 at UHF is out of phase with the bottom λ/4. This means cancellation occurs in the radiation pattern and the majority of the energy is launched at a take-off angle of 45°. This results in about a 4 to 6dB loss in the horizontal plane compared to a conventional λ/4 vertical placed over a ground plane. A horizontal radiation pattern obtained from *EZNEC* **[1]** is shown in **Fig 5.23**. Notice that the 3λ/4 radiator has most of its energy at 44°.

Thus, although an antenna can be made to work at its third harmonic, its performance is poor. What we need is a simple, reliable method to decouple the remaining λ/2 at UHF of a 2m radiator, but have it remain electrically unaffected at VHF. We want independent λ/2 radiators at both VHF and UHF frequencies. The original DBJ-1 used a combination of coaxial stubs and 300Ω twinlead cable, as shown in **Fig 5.24**. Refer to **Fig 5.24**, and start from the left hand bottom. Proceed vertically to the RG-174A lead in cable. To connect to the antenna, about 5 feet of RG-174A was used with a BNC connector on the other end. The λ/4 VHF impedance transformer is made from 300Ω twin lead. Its approximate length is 15 inches due to the velocity factor of the 300Ω material. The λ/4 piece is shorted at the bottom and thus is an open circuit (high impedance) at the end of the λ/4 section. This matches well to the λ/2 radiator for VHF. The 50Ω tap is about 1.25 inches from the short, as mentioned before.

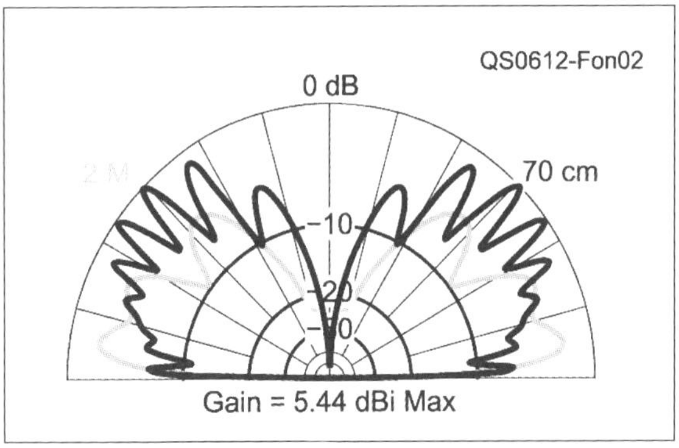

Fig 5.23: Elevation plane pattern comparing 2m J-pole on fundamental and on third harmonic frequency (70cm), with the antenna mounted 8ft above ground. Most of the energy at the third harmonic is launched at 44°.

For UHF operation, the λ/4 matching stub at VHF is now a 3λ/4 matching stub. This is electrically a λ/4 stub with an additional λ/2 in series. Since the purpose of the matching stub is for impedance matching and not for radiation, it does not directly affect the radiation efficiency of the antenna. It does, however, suffer some transmission loss from the additional λ/2, which would not be needed if it were not for the dual band operation. I estimate this loss at about 0.1dB. Next comes the λ/2 radiating element for UHF, which is about 12 inches. To make it electrically terminate at 12 inches, a λ/4 shorted stub at UHF is constructed using RG-174A. The open end is then connected to the end of the 12 inches of 300Ω twinlead.

The open circuit of this λ/4 coax is only valid at UHF. Note that it is 4.5 inches and not 6 inches due to the velocity factor of RG-174A, which is about 0.6. At the shorted end of the 4.5-inch RG-174A is the final 18 inches of 300Ω twinlead. Thus the 12 inches for the UHF λ/2, the 4.5 inches of RG-174A for the decoupling stub at UHF, and the 18 inches of twinlead provide for the λ/2 at 2m. The total does not add up to a full 36 inches that you might think. This is because the λ/4 UHF RG-174A shorted stub is inductive at 2m, thus slightly shortening the antenna.

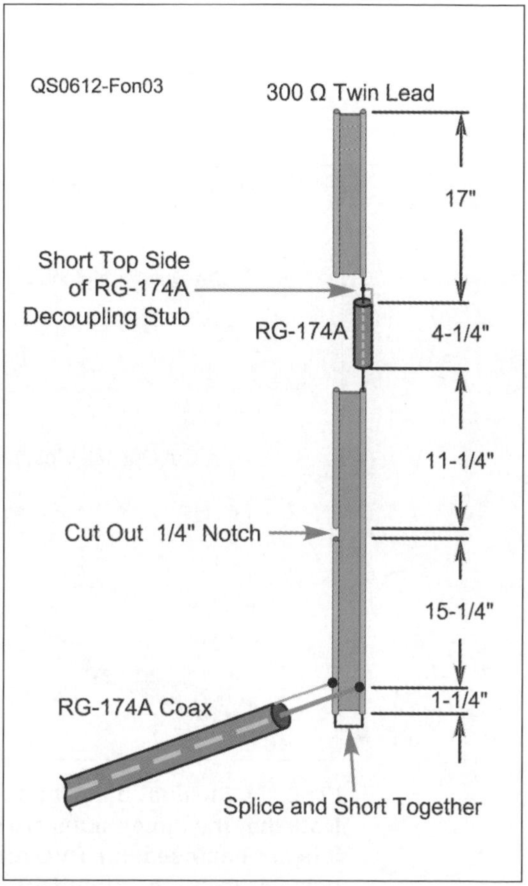

Fig 5.24: The original DBJ-1 dual-band J-pole (these dimensions assume the antenna is inserted into PVC pipe).

The original DBJ-1 had the antenna inserted into a 6ft long PVC pipe: fine for fixed operation but not really suitable for portable use. The new antenna had to have the ability to be rolled up when not in use and had to be durable enough for use in emergency communications. The challenge was to transfer the concepts developed for the DBJ-1 and apply them to a durable roll-up portable antenna. After much thought and experimenting, I adopted the configuration shown in **Fig 5.25**.

The major challenge was keeping the electrical characteristics the same as the original DBJ-1 but physically constructing it from a continuous piece of 300Ω twinlead. Any full splices on the twinlead would compromise the durability, so to electrically disconnect sections of the twinlead, I cut small 1/4-inch notches to achieve the proper resonances. I left the insulating backbone of the 300Ω twinlead fully intact. I determined the two notches close to the λ/4 UHF decoupling stub by experiment to give the best SWR and bandwidth.

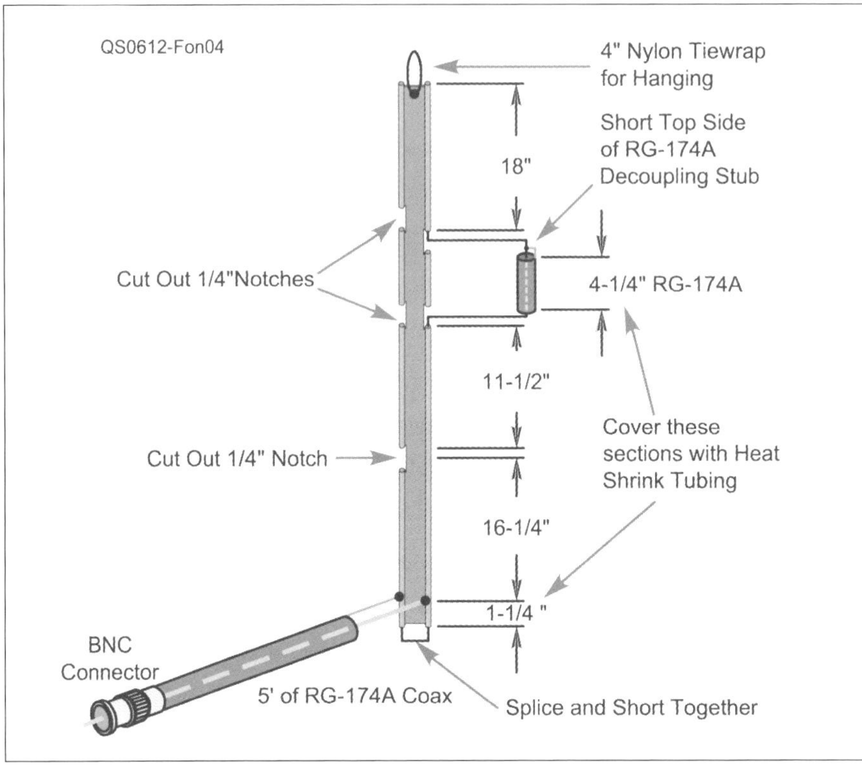

Fig 5.25: The dual-band J-pole, modified for portable operation (the 'DBJ-2'). Note that the dimensions are slightly longer than those in Fig 5.24 because it is *not* enclosed in a PVC tube. Please remember that the exact dimensions vary with the manufacturer of the 300Ω line, especially the exact tap point where the RG-174A feed coax for the radio is connected.

Because this antenna does not sit inside a dielectric PVC tube, the dimensions are about 5% longer than the original DBJ-1. I used heat shrink tubing to cover and protect the UHF $\lambda/4$ decoupling stub and the four 1/4-inch notches. Similarly, I protected with heat shrink tubing the RG-174A coax interface to the 300Ω twinlead. I also attached a small Teflon tie strap to the top of the antenna so that it may be conveniently attached to a non-conductive support string.

The photo opposite shows a picture of the $\lambda/4$ UHF matching stub inside the heat shrink tubing.

The DBJ-2 can easily fit inside a pouch or a large pocket. It is far less complex than what would be needed for a single band ground plane, yet this antenna will consistently outperform a ground plane using three or four radials. Set-up time is less than a minute. As with any antenna, it works best as high as possible and in the clear. To hoist the antenna, use non-conducting string or fishing line.

The top of the DBJ-2 is a high impedance point, so objects (even if they are non-metallic) must be as far away as possible for best performance.

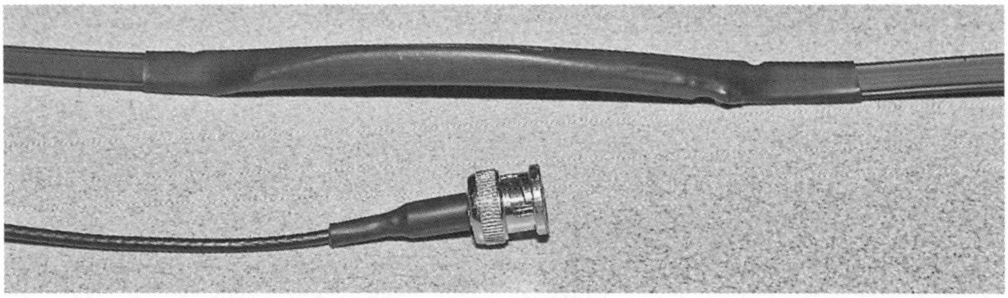

The λ/4 UHF RG-174A decoupling stub covered with heat shrink tubing. This is shown next to the BNC connector that goes to the transceiver.

The other sensitive points are the open end of the λ/4 VHF matching section and the open end of the λ/4 UHF decoupling stub.

I measured the DBJ-2 in an open field using an Advantest R3361 Spectrum Analyser. The results are shown in **Table 5.4**. The antenna gives a 7dB improvement over a flexible antenna at VHF. In actual practice, since the antenna can be mounted higher than the flexible antenna at the end of your handheld, results of +10dB are not uncommon. This is the electrical equivalent of giving a 4W handheld a boost to 40W.

The DBJ-2 performs as predicted on 2m. It basically has the same performance as a single band J-pole, which gives about a 1dB improvement over a λ/4 ground plane antenna. There is no measurable degradation in performance by incorporating the UHF capability into a conventional J-pole.

The DBJ-2's improved performance is apparent at UHF, where it outperforms the single band 2m J-pole operating at UHF by about 6dB: this is *significant*. See **Table 5.5**. I have confidence in these measurements since the flexible antenna is about –6dB from that of the λ/4 ground plane antenna, which agrees well with the literature.

Also notice that at UHF, the loss for the flex antenna is only 2.0dB, compared to the ground plane. This is because the flexible antenna at UHF is already 6 inches long, which is a quarter wave. So the major difference for the flexible antenna at UHF is the lack of ground radials.

The DBJ-2 antenna is easy to construct, low cost and very compact. It offers significant improvement in both the VHF and UHF bands compared to the stock flexible antenna antenna included with a handheld transceiver.

VHF λ/4 GP 4 radials	Flexible Antenna	Standard VHF J-Pole	Dual-Band J-Pole
0dB reference	–5.9dB	+1.2dB	+1.2dB

Table 5.4: Measured relative performance of the dual-band antenna on 2m.

UHF λ/4 GP 4 radials	Flexible Antenna	Standard VHF J-Pole	Dual-Band J-Pole
0dB reference	–2.0dB	–5.5dB	+0.5dB

Table 5.5: Measured relative performance of the dual-band antenna on 70cm.

A UNIQUELY TUNED 2M ARDF LOOP

John Portune, W6NBC
QST, January 2013

One of the hidden joys of ARDF (amateur radio direction finding – transmitter hunting) is that it provides an excuse for 'homebrewing'. As a new participant in this radio sport, John, W6NBC, found himself constructing antennas he would never have considered for other activities. This little loop is a perfect example.

Left to right, loop, coupling loop, balun, attenuator and handheld transceiver mount (behind attenuator).

At the starting point of a transmitter hunt, one needs high gain – a Yagi or quad antenna, and a sensitive receiver. As you approach the hidden transmitter, however, the game changes dramatically. Now you need a portable antenna and an attenuator. This is the realm of the loop or a hand-held Yagi. I prefer the loop. It's handier and, in most cases, just as effective.

Compared to a Yagi or a quad, a small loop has downsides, even though it does have equally good directivity. One is its size, which does not let it capture as much signal, particularly if not tuned to the frequency of the transmitter. Commercially manufactured direction finding (DF) loops are often not tuned. They are designed this way to permit them to be used on more than one band.

I preferred to increase my loop's ability to receive by tuning it to 2m with a variable capacitor at the loop's ends. The loop's untuned resonance is roughly 450MHz, making it inefficient on 2m. For locating radio-tagged wild animals, finding memory impaired patients or locating a VHF aviation beacon, it is easy to retune the capacitor for another VHF band.

I chose a somewhat unconventional variable capacitor simply because common rotary and compression mica variables are awkward to add to a loop. Furthermore, they are becoming hard to find. At first I did not know what kind of variable capacitor to use. Then lightning struck. Why not dispense with using a discrete capacitor at all? Let the loop itself be one of the plates of a variable capacitor by constructing it of tubing instead of solid rod or wire. The other plate of the capacitor is 11.5 inches (29cm) of the centre conductor and dielectric of a piece of RG-8X coax, with the black outer

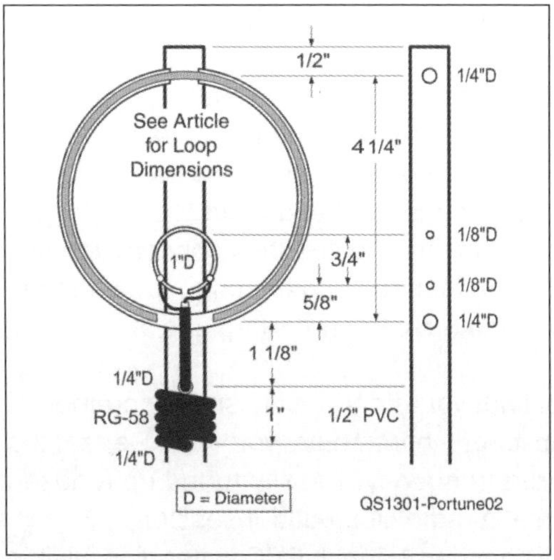

Fig 5.26: The transmitter hunting loop. Note that the tuning capacitor is formed by the tubing of the loop itself and an inserted length of the inner conductor and dielectric from RG-8X coax.

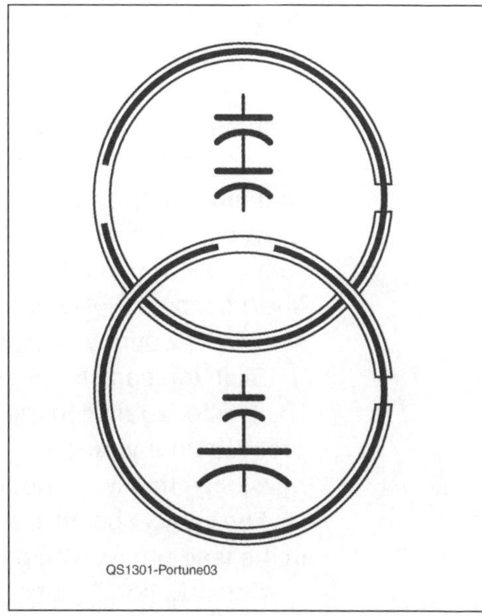

Fig 5.27: Cross sections of the loop with coax inner conductor and dielectric from RG-8X coax in two different positions. Maximum capacitance is with the inner conductor as in the top loop.

jacket and shield removed. It's inserted inside the tubing of the loop as shown in **Fig 5.26**.

Notice that this design creates two capacitors in series. One is above the gap, the other is below it. As illustrated in **Fig 5.27**, when the sides are equal total capacitance is maximum. Move the sliding piece to either side and the capacitance decreases, increasing the resonant frequency. Note the equivalent circuit of the two capacitors. Therefore, to adjust the frequency of the loop you merely slide the coax inner conductor and dielectric with your fingernail at the gap in the top end of the boom (see photo).

Loop from the top end. Tune the loop by sliding the coax inner conductor and dielectric with your fingernail.

The tuning is very stable. I've never needed to retune my loop, though I did have to experiment, using an antenna analyser, with tubing size and inner conductor length to find the right combination. Just duplicate the dimensions given in **Fig 5.26** and you will have no trouble tuning the loop.

There are several possible ways to match the loop to 50Ω coax. I use compact and flexible RG-58 for this application. From practical experience with small loops, I have found that an inductive coupling loop is the easiest matching method. This one is 1 inch in diameter, made from #14 AWG solid copper wire. Note that there is a gap in it inside the boom. Solder the feed coax to the coupling loop just outside of the boom. There is no electrical connection to the main loop. The two constitute an RF transformer with energy passing via magnetic induction.

The easiest way to tune up the antenna is with an antenna analyser such as an MFJ-259 or equivalent. Begin by centring the coax inner conductor. Next find the resonant frequency by sweeping across the VHF band with the antenna analyser connected to the feed coax and looking for the SWR dip. It will be below the 2m band at first. Again, don't concern yourself with SWR. If you later want to use the loop for transmitting, you can then fine tune the coupling loop for a better match. For receiving, the dimensions shown are entirely adequate.

Slide the coax inner conductor with your fingernail in small increments. Each time, again find the dip. Gap length increments work fine. Repeat the process until you reach the desired frequency. I easily tuned up a dozen new loops this way one afternoon at a radio club build-it session.

The vertical boom is a length of common 1/2 inch PVC water pipe, visible in the lead photo. The pipe shown is actually the handle of a dust pan.

Construction is very straightforward. Cut and bend the loops with your fingers. A form is not necessary. Then drill the holes in the boom for the loops and balun. Next, rotate the loops into place through the holes and add the coax, forming the balun as you go. Finally, solder the ends of the coax to the coupling loop and attach a connector.

Carol Adams, KG6VLG, demonstrating the use of *body fade*. Note the placement and orientation of the loop. She is facing the hidden transmitter.

To use the loop as part of the sport of radio transmitter hunting you will need an attenuator. It cuts down the strong signals as you get close to the hidden transmitter. Without it, your receiver will saturate and you will lose the ability to determine directions through peaks and nulls. Two types of attenuator are common. One operates with the receiver 'on channel' and attenuates with switchable resistive pads or a potentiometer. The other, called an off-set attenuator, mixes the incoming signal with a local crystal oscillator, typically at 2 or 4MHz. Hence, you listen high or low. With a resistive on-channel attenuator your receiver needs to be shielded to prevent direct pick-up through the case. An off-set attenuator is much simpler to use.

When using this antenna, keep in mind that there are two nulls and two peaks. The peaks occur when the loop is in line with the hidden transmitter, the nulls when it is broadside. To resolve this all you need is a very elementary transmitter hunting technique called *body fade* (see photo). With the loop held close to your chest, and also in line with the transmitter, turn in a full circle. You will still get two peaks (and two nulls). But with your body present the correct and stronger peak will occur when you are *facing* the hidden transmitter. Once you know the right direction, hold the loop high above your head. The peaks and nulls will be more evident.

Also essential to all transmitter hunt antennas is a balun. If it is omitted, the antenna pattern will be skewed, which could lead you in the wrong direction. Drill four additional holes in the boom to secure it. This type of balun is an RF choke formed of five turns of the feed coax. VHF ferrite beads may also be used.

A PORTABLE TWO-ELEMENT 6M QUAD

Pete Rimmel, N8PR

QST, June 2014

Perfect for VHF Field Day, or any other 50MHz portable operation, this portable quad antenna made from easily-available PVC tubing can be assembled or disassembled in less than five minutes.

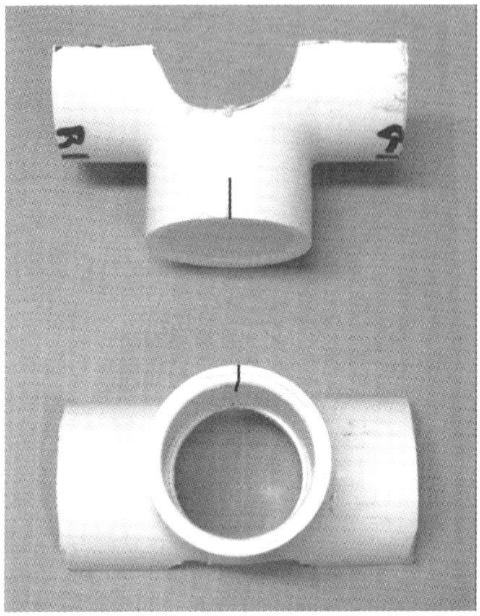

Split a PVC T and file away material so that with a four-way PVC cross you have a fixture with five openings.

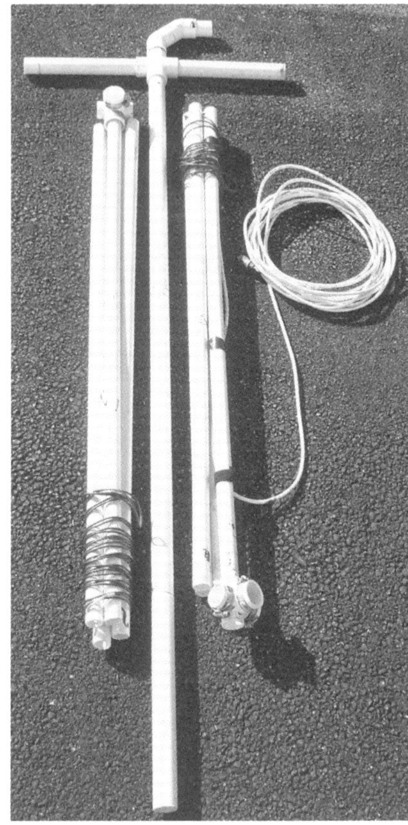

Right: Easy-to-stow disassembled antenna components.

The author of this article, Pete, N8PR, wrote that he devised a unique way to lock the spreaders into the centre PVC crosses using the antenna wires of the quad. This key feature allowed him to tighten the wires so that they won't sag while in use, but lets them be loosened for disassembly and storage in a small space such as a car.

You can construct this quad from parts available at any DIY or 'home improvement' store. The PVC dimensions are not critical, and can be within 1/4 inch. The centre crosses of 3/4-inch pipe are compatible with 1-inch 'T's that were split to create the boom connections. The inner diameter of the 1-inch T matches the outer diameter of the 3/4-inch crosses with a bit of trimming or filing. This larger size PVC is more rigid and makes the antenna more stable.

#14 AWG stranded insulated wire was used for the antenna elements: insulated wire is less troublesome to store and won't snag or break while in storage.

Cut four 40-inch (102cm) pieces of schedule 20 × 3/4-inch PVC for the driven element spreaders, and four 41.25-inch (105cm) pieces for the reflector spreaders. From a 10-foot (3.05m) section of schedule 20 × 1 inch PVC, cut two 11-inch (28cm) pieces for the boom. Use the rest for the vertical mast.

You will also need two 3/4-inch PVC crosses (these have four openings); two 1-inch PVC Ts (these have three openings, see the photo opposite), one 1-inch PVC cross and six 1.75-inch (5cm) diameter stainless steel hose clamps. Cut an 18' 10.5" (5.75m) length of insulated #14 AWG stranded wire for the driven element and a 19' 5.75" (5.94m) length for the reflector.

You'll also need three 2-inch long pieces of 3/8-inch heat shrink tubing, coax sealant, black electrical tape, Velcro or similar hook-and-loop fastener strips, and a length of 50Ω coax to reach to your station.

Cut two pieces of the 1-inch PVC pipe to 11 inches (28cm) long. Place the 11-inch long 1-inch diameter PVC tubes into the 1-inch cross which also connects to the boom, as seen in the lead photo. The rest of the 1-inch PVC goes into the lower part of the 'T' to form the mast.

Make a pair of 'five opening fixtures' for connecting the four cross pieces to each end of the boom, as follows. Make two 'half-Ts' by cutting the long side of a 1-inch 'T' with a hacksaw as shown in the photo opposite. Cut a slit in the round part of this fitting. Shape the cut area, and remove a half circle of material using a rasp or file so that it will fit snugly over the 3/4-inch cross. Use two hose clamps to fit each half-T to each PVC cross (see the photo below left). The hose clamps should pull the 'T' snug to the smaller cross without deforming it. A third hose clamp compresses the slit and holds the element on the boom.

Drill holes for the wires 1/2 inch (about 1.25cm) from the end of each of the eight spreader pieces, going through both walls of the pipe. Take care to centre these holes so the wire will go straight through the centre of the spreader. Cut J-shaped openings in three of the 40-inch long spreaders and all four of the 41.25-inch spreaders. Following the progression from left to right in the photo below right. Drill a second hole in each spreader at about a 30° rotation and about 1.25 inches in from the end. Again, be sure to drill straight through both walls of the pipe. Draw some guidelines on each of the seven pieces of PVC. Drill a series of

The five-opening fixture ties the boom end to the four spreaders.

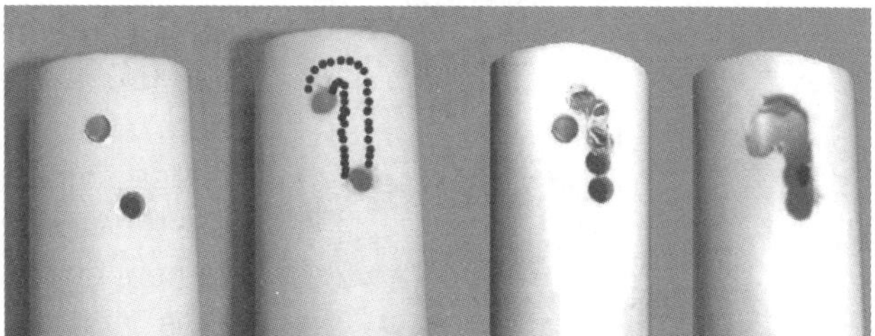

From left to right, the steps required for making the 'J holes'.

holes to form an upside down J on both sides of each tube. Be sure to leave PVC in the gap between the holes on the right and the single hole on the left so that the wire can lock in two positions. Carefully 'connect the dots' with a drill, and remove all the material between the holes. This forms the unique adjustable J feature of the wire guide on seven spreaders.

Strip 1 inch of insulation from each end of the longer #14 AWG wire and twist the stranded wire to keep it neat. Thread the wire through the eight J holes of the four longer spreaders. Place a 2-inch piece of heat shrink tubing over one end of the wire. Overlap the two ends by 1 inch, twist and solder them together.

Slide the heat shrink over the soldered connection and heat it to waterproof the connection. You now have a loop with the wire going through the four outer ends of the spreaders. Position the wire in the inner part of the J on all four spreaders. Insert the four spreaders into one of the two 3/4-inch crosses. After you seat the pieces snugly into the crosses, move the wire to the outer part of the J to tension the wire element. You might not need to tension the wire on all four spreaders. Twist the pipes so that the holes line up and the wire lies in a flat loop when assembled. Your reflector is now completed.

For the driven element strip 1 inch of insulation from each end of the shorter length of wire. Prepare a length of 50Ω coax (e.g. RG8-X, RG-58 or RG-213) by stripping the outer jacket back about 1.5 inches. Comb the braid out so that it can be separated from the centre conductor and twist the braid into a single strand. Cut it to a 1-inch length. Strip 3/4 inch of the insulation away from the centre conductor. These dimensions contribute to the overall length of the driven element.

Thread the insulated #14 AWG wire through the holes in the four shorter (40-inch) spreaders. Be sure that the bare ends of the wire are near the spreader without the J cut into the end. This is where you connect the coax. Slide a 2-inch piece of shrink tube over each end of the #14 AWG wire.

Carefully wrap the centre conductor of the coax to one stripped end of wire and solder it. Do not overheat. Wrap the other wire around the braid of the coax and solder it in place. Slide the two pieces of shrink wrap over the soldered connections and heat them. Seal the remaining area of the coax braid with a small amount of coax sealant and cover with black tape to completely waterproof the connection (see photo).

Detail of the coax feed line connection.

Use some loops of Velcro (or black tape) to secure the coax to the PVC tubes. Assemble the driven element by placing the four spreaders into a 3/4-inch cross, and place the wires in the J cuts to tension the wire and rotate so that the loop is flat.

It is easy to assemble and disassemble the antenna in the field. First, separate the two elements and lay the reflector parts on the ground. Be sure that the wire is in the long part of the J holes. Place the four elements into the cross. Move the wires into the shorter part of the J holes to tension the wire and keep the spreaders from dropping out of the cross. Slide the director on to the boom and partially tighten the hose clamp over the slit in the 'half-T' to secure the element to the boom. Assemble the driven element as you did the reflector and mount it on to the boom using another hose clamp. Be sure to orient the elements in the polarisation you want before tightening the hose clamps completely. Route the coax along the element and boom to the mast using Velcro loops to keep it in place.

When you take the antenna apart, leave the cross piece attached to one of the four spreaders for storage. Wrap the element wires around the four spreaders to keep them in a neat package.

You can easily choose the polarisation you want. For vertical polarisation, place the driven element so that the coax is on a horizontal spreader, and the antenna is in a 'diamond' configuration. For horizontal polarisation, place the diamond shape so that the coax is at the bottom of the element. You can also orient the elements for 45° polarisation.

This antenna is designed to resonate at 50.1MHz. The VSWR should be less than 1.5:1 from 50.0 to 50.6MHz. The element spacing is designed to give a 50Ω match.

A 6M PORTABLE YAGI ANTENNA

H Scott McCann, W3MEO
QST, April 2018

From a 6m quad to a 6m Yagi. This one breaks down for storage inside its own PVC boom. It uses a painter's pole mast and two telescoping aluminium elements, and is suitable for portable low-power operation, especially in the portable category of VHF contests.

All dimensions are shown in **Fig 5.28**. The boom, made from PVC pipe, also stores the aluminium elements, feed-point adapter, and screwdriver

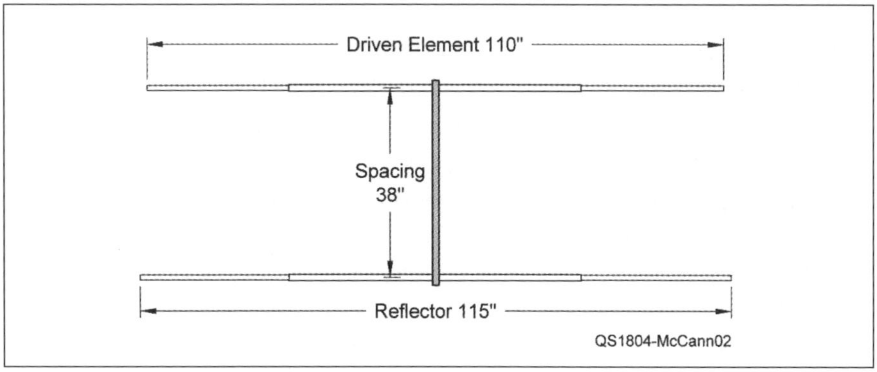

Fig 5.28: Overall dimensions of the antenna.

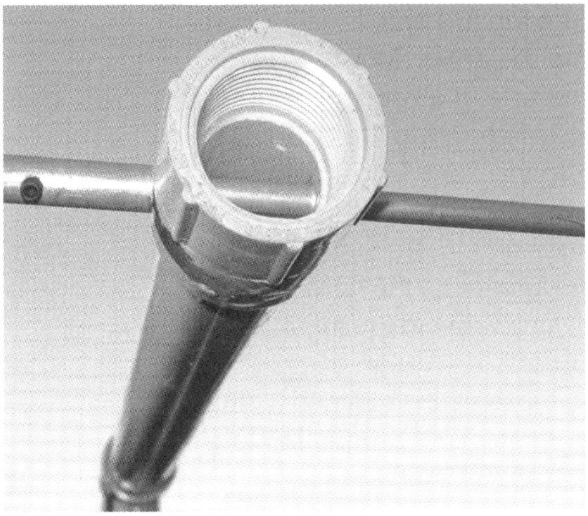

Detail of reflector centre. The threaded piece is the end of the boom. The boom is a piece of PVC pipe which also serves as a storage case. When you take the antenna apart, all parts slide into the boom through the standard PVC to the threaded adapter, and a threaded PVC pipe cap screws into the hole to close the case / boom. The cap is removed in the photo to show the construction of the end of the antenna.

style Allen wrench needed for assembly. The feed-point adapter is fashioned from a piece of brass.

The boom is made from pieces of 1-inch PVC pipe joined in the middle with a 1-inch to 1/2-inch 'T'. The boom length is such that the two elements are 38 inches (96.5cm) apart when assembled. The T screws on to a painter's pole that acts as a mast. A pipe cap is cemented to one end, while the other end has a screw cap adapter to close the tube when it is used as the carrying case (see photo, left).

Drill the 1/4-inch holes for the inner element sections through the caps for solid support. It is very important to do a lot of dry fitting before cementing, as PVC cement sets nearly instantly, allowing no time for adjustment. It has to fit together properly the first time.

Both elements are made from 3/8-inch aluminium tubing cut 32 inches (81cm) long for the inner portions, and 1/4-inch aluminium tubing for the outer portion, cut as described in **Table 5.6**.

PVC parts, from your local D-i-Y / hardware store
1-inch PVC pipe cut to length (see text)
Slip-on PVC cap (see photo above)
Slip-on cap with threaded opening and screw-in plug (see photo above)
1-inch to 1/2-inch threaded T
1 can of PVC cement
Aluminium parts (e.g. from www.aircraftspruce.com)
(4) 32-inch inner portions, 3/8 × .049 AS# 03-39550-6
(2) 27.5-inch reflector outer portions, 1/4 × .035 AS# 03-39300-6
(2) 24.5-inch driven element outer portions, 1/4 × .035 AS# 03-39300-6
(1) 8-inch reflector centre rod, 03-45600-2 AS#0345600-1
(10) 8-32 × 1/4 Allen screws, AS#04-00183
Fibreglass part (e.g. from intothewind.com)
(1) 1-foot length of 1/4-inch fibreglass rod

Table 5.6: Materials required for the 6m portable Yagi.

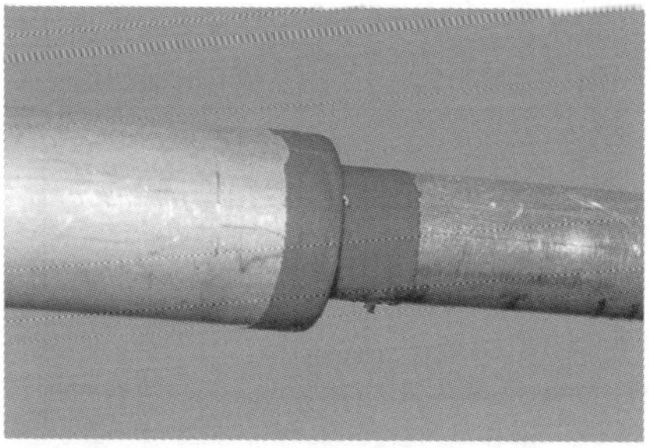

Paint marks on driven element showing proper extension distance. See the text for dimensions.

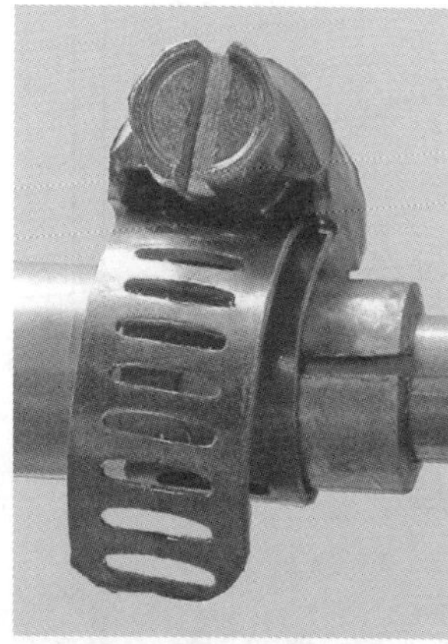

An alternate method of element construction using small hose clamps.

The central portion of the reflector uses a quarter-inch aluminium tube.

The central portion of the driven element is a 1/4-inch fibreglass rod. Inner portions are drilled and tapped, for 3/16in-long 8 × 32 Allen screws, about 1 inch from each end to secure to the centre and outer portions. Outer portions are adjusted for overall element length and marked with paint (see photo above left) for easy assembly in the field without further measuring. Alternatively, slit the 3/8-inch pipe and clamp with small hose clamps (see photo above right).

One of the driven element inner portions is threaded 3/8in × 24 to attach the feed point (see photo below). The feed-point adapter, as shown in the photo, is a piece of brass tapped at one end for one side of the driven element, with a 3/8-inch hole at the bottom for a BNC connector. The centre terminal of the BNC attaches to the piece of brass or copper tube with a 2.5-inch piece of wire. The tube is large enough to slip on the other side of the driven element, secured by an 8-32 Allen screw. Connect the coax with a right-angle BNC adapter.

Detail of feed-line connection to driven element. (No current choke or balun was used at the feed point.)

Fig 5.29 shows the azimuthal pattern of the antenna, and **Fig 5.30** shows the SWR from 50 to 50.5MHz, both simulated using *EZNEC* software. The theoretical match was confirmed on the air; my rig showed about 1.5:1 SWR,

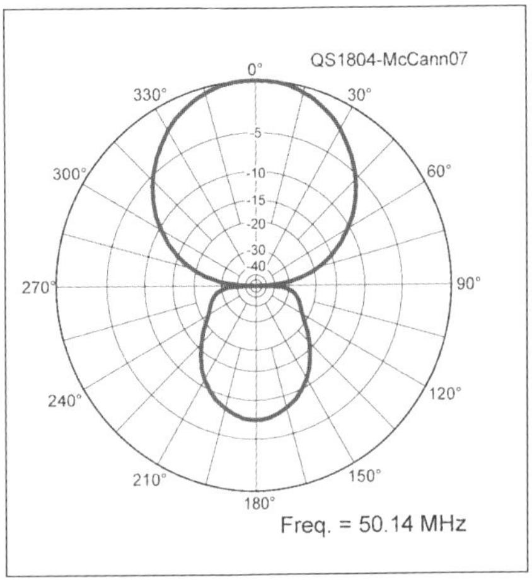

Fig 5.29: Horizontal (azimuthal) plot simulated using *EZNEC*.

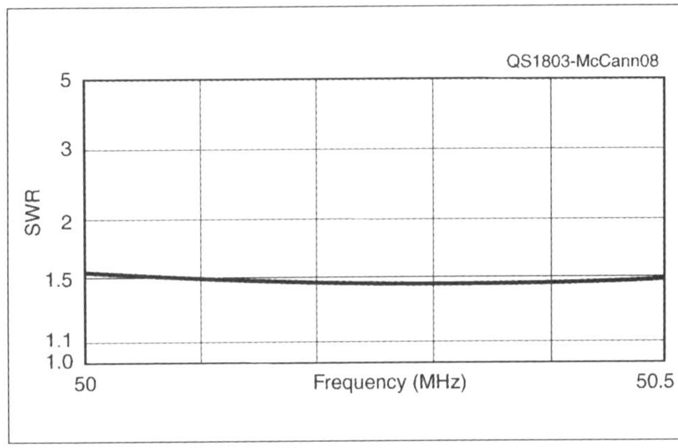

Fig 5.30: SWR plot simulated using *EZNEC*.

meeting my expectations. This beam is not designed for permanent installation, but it has stood up to brief summer storms. Using this antenna, I added some 'awards wallpaper' to my shack, and I hope it does the same for others who try it.

The completed 2-element 6m portable Yagi.

SOME IDEAS FOR QUICK DEPLOYMENT VHF / UHF ANTENNAS, INCLUDING 'THE SELFIE-STICK YAGI'

Kevin Richardson, G0PEK
RadCom, November 2017

Many people have a selfie-stick hanging around. Kevin, G0PEK, wrote "I have found a much cooler use than taking blurry 'I'm-the-centre-of-the-universe' photos: turn it into an amateur radio aerial!" Before describing the

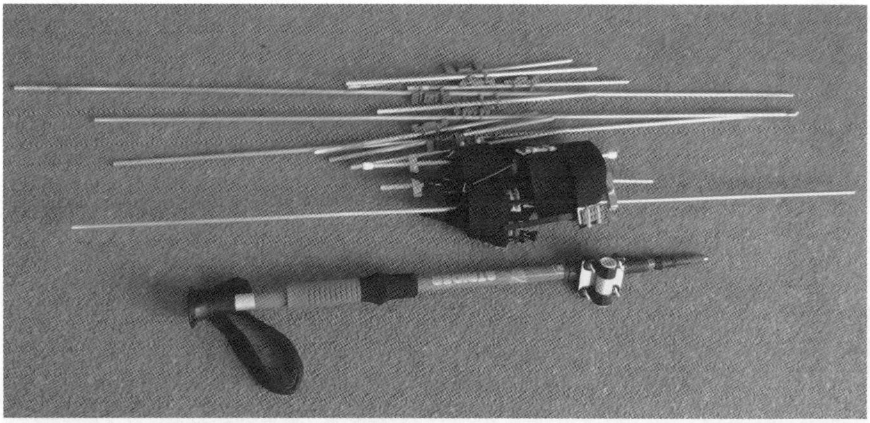

Dual band roll-up Yagi, ready for transport. The walking pole in the foreground shows the mounting bracket

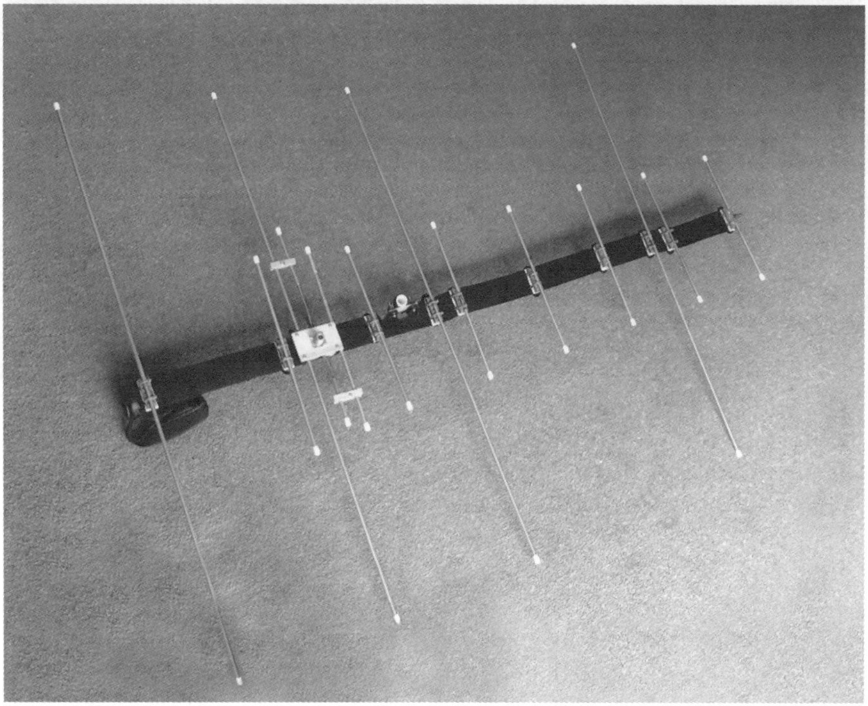

Roll-up Yagi attached to a walking pole and ready for use.

'Selfie-Stick Yagi', though, Kevin discusses some of his earlier experiments with small directional antennas that he used for his summit and portable operating. He wrote:

Two years ago, I tried a roll-up Yagi. This was achieved by mounting the elements on a length of webbing tape. A gamma match was employed for tuning. I also found that it was possible to mount a 144MHz and 432MHz beam on the same piece of tape and employ one feed to a pair of gamma

Lightweight Moxon beam mounted on a guyed fibreglass fishing rod atop Red Crees.

matches, thereby making a rather useful 4-element VHF and 8-element UHF antenna.

To deploy it, the elements were attached to a walking pole with releasable zip ties. A small bracket bolted to the walking pole allowed it to be mounted horizontally or vertically. Some very good contacts were made with this antenna. The elements were made from old archery arrows, which made for a robust design when deployed, but it was a little time consuming to set up.

My next 2m antenna was a simple Moxon. It used lightweight plastic pipes for the spreaders. This collapses down nicely for transport and is very quick to deploy. It became our antenna of choice on lightweight activations, such as that shown in the photo above.

But what I really wanted was something for the occasional portable activation and to pick up signals from the ISS. Ideally this would be a beam that fitted in a pocket of my rucksack.

In case you haven't come across selfie-sticks, they have much in common with the telescopic antenna rods found, for example, on broadcast VHF / FM portable radios (remember them?) My daughter, Lauren, M6HLR, had a couple of selfie-sticks given to her at Christmas. She gave one to me and I wondered if I could do anything constructive with it. The length looked useful so I started to see if it was possible to make a beam antenna with one.

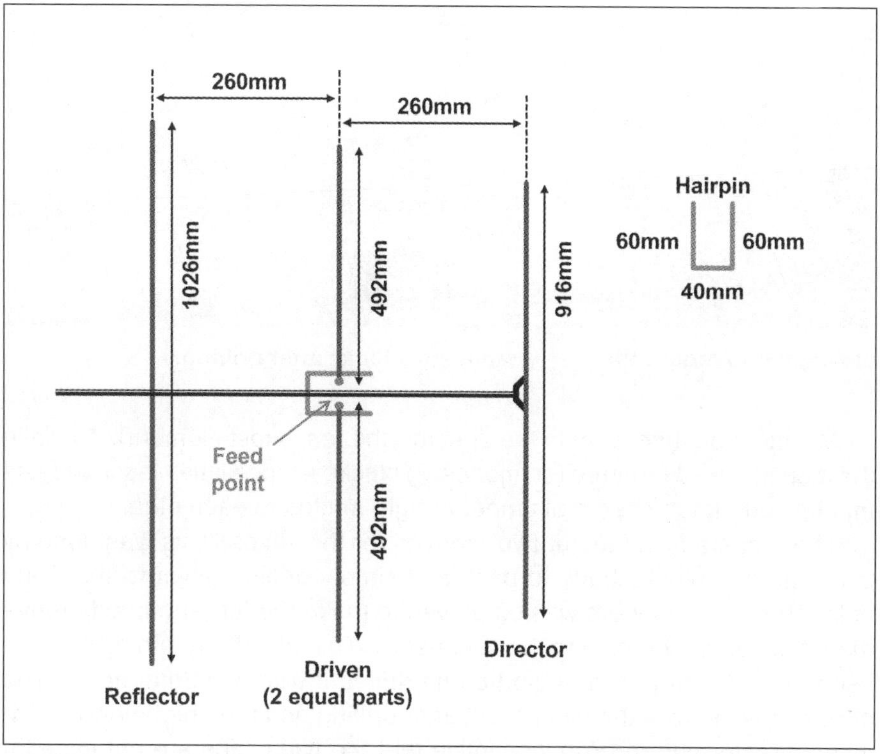

Fig 5.31: General assembly of the Selfie-Stick Yagi. Note that these dimensions are for guidance only – see text.

The result was the 'Selfie-Stick Yagi' and I'll describe the various steps I took to create it. Note, however, that the dimensions I've listed here are all specific to the bits and pieces I used. The materials you find to build yours will be different and so will be the exact dimensions. It's easy enough to find Yagi dimension calculators online and plug in the sizes of whatever you've got to hand. Go with what the calculator says: only worry if it gives you lengths markedly different from mine (e.g. a reflector of about 1500mm instead of about 1000mm).

Fig 5.31 shows the general layout and dimensions of my prototype.

I purchased 6 x 50cm telescopic whips for the elements (making seven in total), intending to make a 3-element Yagi. The rest of the components came from the shack and DiY stuff in the garage. Nothing is particularly special and you can merrily substitute items – just remember to put the dimensions into the online calculator.

The selfie-sticks each came with a bracket to hold the phone. When I removed that, it left a C-shaped brace, ideal for mounting the director (front-most) element. I carefully drilled out the bolt so that two whip elements could be slotted through and bolted together with an M3 nut and bolt. I found that drilling a small hole directly down the middle of the bracket allowed the M3 bolt to slot in, thereby stopping the element from twisting.

C-shaped bracket adapted to mount the reflector.

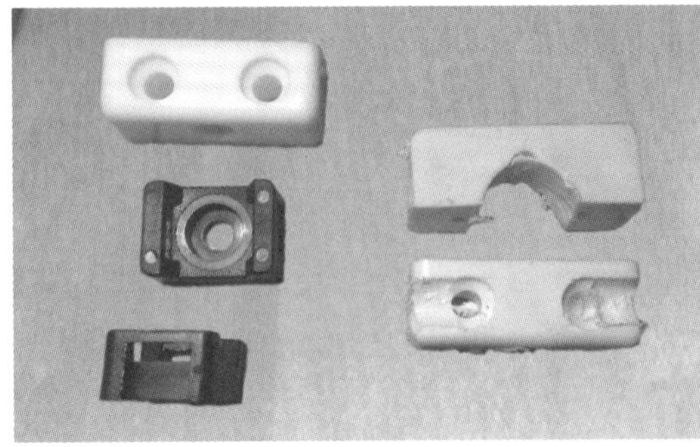

'Modesty' blocks after drilling.

My attention then went to the director (the rear-most element). To make the mount, I used furniture (or 'modesty') blocks – those little plastic widgets that let you attach sheets of timber at right angles to each other.

The first step was to cut two grooves for the whips. This was done by clamping two blocks back to back and simply drilling down to the fixing hole. The same concept was employed to make the larger curved groove that attaches the bracket to the selfie-stick. To make the retaining clamp, I used a plastic strip from a zip tie and drilled it with two 3mm holes. The photo above shows the blocks just after drilling and the one below the finished director centre. You will notice that the two whips are not joined. I used a bridging wire to connect the elements together, thereby increasing the overall element length by 20mm.

Now on to the driven element. I used a small project box, a bulkhead BNC socket and another modesty block to form the centre. I drilled into the narrow ends of the modesty block and then cut in half lengthways. This left the slots for the elements. Then it was just a matter of drilling a couple of holes in the plastic box for the elements and a bigger hole for the BNC socket.

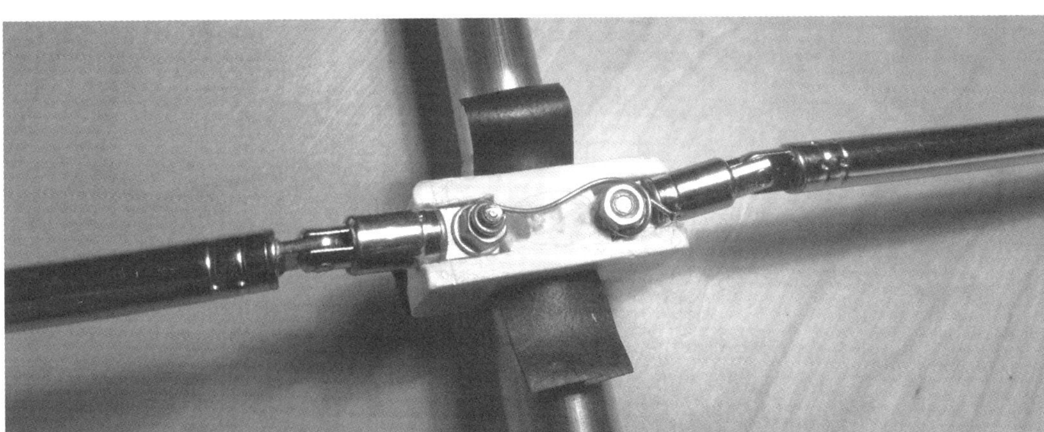

Finished director centre.

Connections between the BNC socket and elements were made using a couple of M3 solder tags attached to small bit of coax. The half section of modesty block was then secured to the plastic box with the same M3 screws that hold the driven element in place. The photo to the right shows the finished thing.

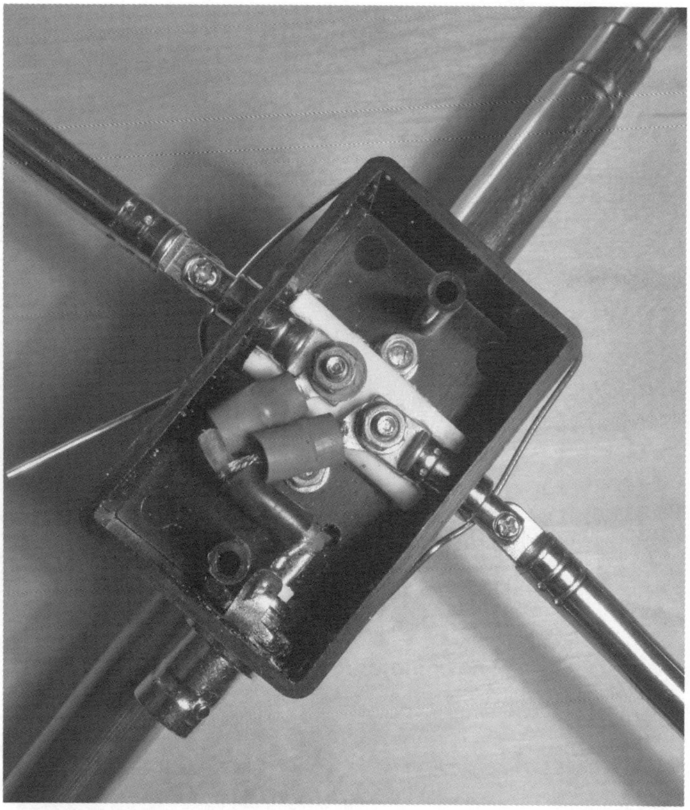

Dipole centre.

I used a couple of zip tie wall mounts bolted to the underside of the box, which made it easy to zip tie the driven element assembly to the selfie-stick.

Using dimensions from whichever Yagi calculator you used, extend the selfie-stick and secure the driven element the correct distance from the reflector. You may find a strip of rubber tape stops the element from sliding on the selfie-stick boom.

I found that when fully extended, the director and driven element were more or less the correct length. The director was a bit long, so I retracted the thinnest section.

To tune the aerial, I used a piece of stainless steel wire left over from another project to make a hairpin match (but any other similar piece of solid wire will do). This was bent into a C shape and slotted into the gaps of the hinge section on the two driven element whips. Experimenting with the length of the hairpin, we were able to tune the antenna. **Fig 5.32** and the photo above will probably help you see what I mean.

I first tested the aerial during a QRP Backpacker Contest by mounting it on our 4m telescopic pole. Perhaps it isn't the most efficient contesting antenna but several 100+ mile contacts were made.

The biggest issue when using the antenna like this is the wind. I constantly had to straighten the elements, as the hinges would fold in the wind. There are probably several solutions for this. I used a couple of Biro pen tubes and a piece of insulation tape across the element hinges, which temporarily stiffened them. However, this isn't really an issue if you are hand holding the antenna.

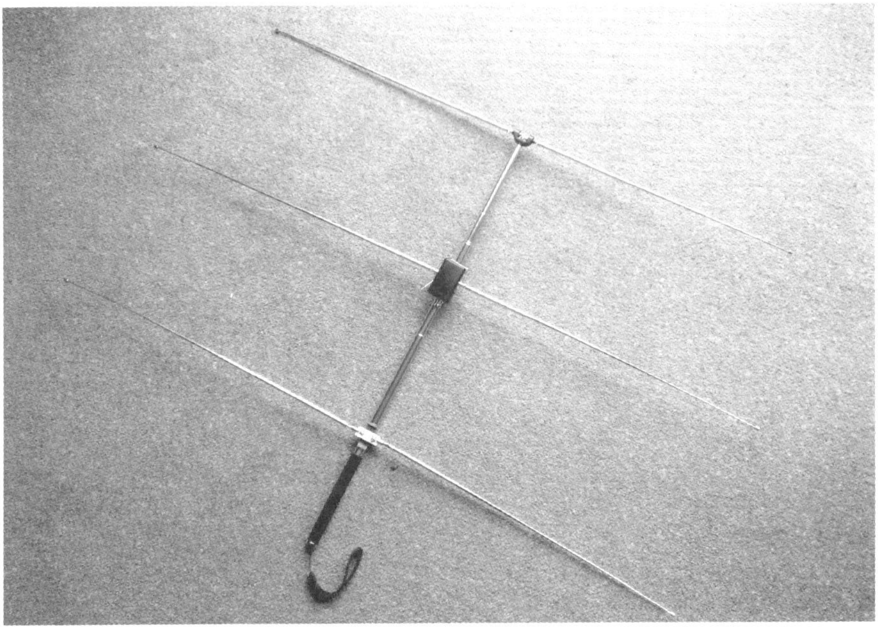

The completed Selfie-Stick Yagi.

The Selfie-Stick Yagi in use, wielded by Lauren, M6HLR.

The end result may not be the most robust antenna around but, for quick activations and portability, you will find it hard to beat. The photo to the left shows me using the aerial – with Lauren, M6HLR, acting as a combined mast and voice-activated rotator.

This is a fun and simple project to make. With a bit of experimentation it should be possible to use the same basic method to create a 70cm version, but I haven't tried it yet. Naturally, all the dimensions will be about one-third those for the 2m version, so the selfie-stick telescopic rods may not be the best choice. However, let the online Yagi calculators be your guide.

REFERENCES

[1] *EZNEC*: www.eznec.com

[2] Kent Britain's Controlled Impedance Antenna: wa5vjb.com/yagi-pdf/cheapyagi.pdf

Chapter 6

Commercial Antennas

THE 'WHIZZ WHIP'

Reviewed by Mike Richards, G4WNC
RadCom, December 2014

The Whizz Whip is a commercial vertical antenna marketed in the UK by Moonraker and intended to allow HF, VHF and UHF portable activity from literally anywhere.

It is a QRP (up to 10W) transmit and receive antenna for 80m to 70cm, with extended receive coverage of 600kHz to 500MHz. As can be seen from the photos, there are two parts to the Whizz Whip, a 130cm telescopic whip antenna and the main control unit. The included instructions comprised a single A4 sheet with the specifications and a few operational instructions.

The Whizz Whip has a BNC connector at the top for connection to the telescopic whip and a PL-259 plug on the front panel for connection to the transceiver. The normal configuration would be to screw the PL-259 connector directly into the antenna socket. This works very neatly on small portable rigs such as the Yaesu FT-817 but it didn't fit so well on an FT-897D: ideally, you need a rig where the antenna socket is mounted towards the top of the rear panel so you can access the rotary tuning knob on the front of the Whizz Whip. In its simplest form, this is all you need to get operational.

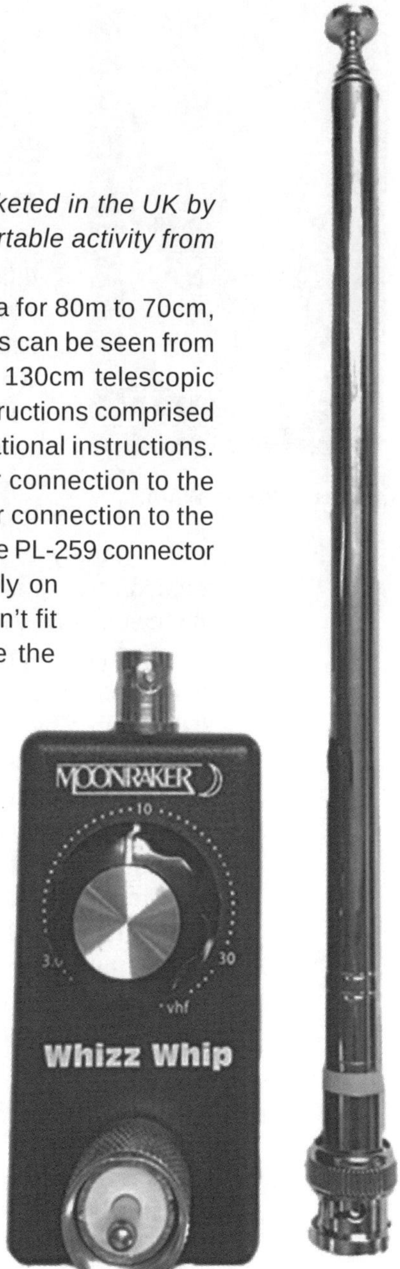

In cases where the Whizz Whip doesn't fit neatly on the back of the rig you can use a coax extension cable. However, there's no mounting system for the control unit so it becomes a bit cumbersome when attached to the whip. One solution could be to mount an SO-239 socket on a vertical metal sheet and attach the sheet to a firm base for stability. That way the Whizz Whip could plug into the new socket where it could then be wired back to the rig. You can also use the Whizz Whip control unit as a universal QRP matching unit to all manner of wire and whip antennas.

The main component in the Whizz Whip is a multi-tapped inductor combined with a rotary switch. The switch has 48 steps and provides access to 47

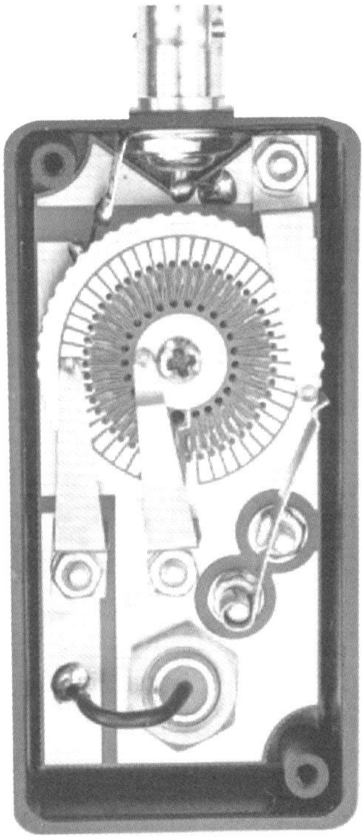

Inside the Whizz Whip.

Band	SWR (Whip only)	SWR (with Counterpoise)
80m	3.6	3.2
40m	1.8	1.6
30m	1.4	1.4
20m	1.6	1.2
17m	1.3	1.3
15m	2.1	2.3
12m	1.7	1.1
10m	1.3	1.5

Table 6.1: Whizz Whip HF band SWR results.

different inductor taps. This would appear to be an L-match arrangement but with a much larger number of taps than usual. The large number of taps is the key to the Whizz Whip being able to work over such a wide frequency range. For VHF / UHF coverage, matching is achieved simply by trimming the length of the telescopic whip element.

Before starting to operate with the Whizz Whip, some SWR measurements were made using a YouKits FG01 antenna analyser. The first set of measurements were taken with only the telescopic whip attached and the Whizz Whip's tuning control adjusted for the lowest SWR. In this configuration, the Whizz Whip is very sensitive to objects in the immediate proximity because the case is made from ABS or similar, so there is no screening. To get reliable measurements the antenna was secured by holding the coax connector in a clamp. For the second set of measurements a 15m-long counterpoise wire attached to the screen of the coax connector was added and run out along the ground. **Table 6.1** shows that it was possible to obtain a workable match on most HF bands. The counterpoise improved the performance on many bands. Obtaining a good match on 80m was extremely difficult as everything was 'hot'.

For the on air tests G4WNC set up a station with the Whizz Whip in his garden, initially just using the telescopic whip. The bands were fairly quiet so he used PSK31 as he said there was almost always activity on 40, 20 or 15m. [Note that this review was carried out in 2014: now in 2020 FT8 or FT4 would be the ideal modes with which to try the Whizz Whip – *Ed.*] The first problem encountered was that RF was getting into his laptop, despite using only 5W of power. To help 'tame' things a 15m counterpoise was connected to the ground connection on the rig and run out across the garden. This improved the situation but there was still some RF getting into the PC: clip-on ferrites connected to the audio leads between the rig and the laptop did the trick. With this arrangement, a few European stations were worked on

PSK31. Best distance on the day was northern Italy: not bad for such a simple setup on a flat band. 599 reports were received, though that's a bit meaningless as everyone gives 599 on PSK: what was more telling was that immediate responses were received to calls, a sure sign that it was putting out a reasonable signal.

Although many will be sceptical of the claims of these 'wonder antennas', this one works, providing the operator takes some care. For HF use it really needs a counterpoise but that is generally easier to set up than a suspended wire antenna. For VHF / UHF operation you simply extend or retract the telescopic whip to achieve a good match.

The Whizz Whip is available from Moonraker (UK) Ltd. See www.moonraker.eu The price is £99.95 inc VAT.

SPIDERBEAM TRIBAND HF YAGI KIT

Reviewed by Rosalie A White, K1STO
QST, July 2005

From the smallest HF antenna covered in this book to the largest. The Spiderbeam may be big but the lightweight 10-15-20m kit reviewed here definitely fits the definition of a portable antenna because it weighs only 12lb (5.5kg) and is delivered in a box just 4 feet (1.22m) long. Heavy duty varieties of the Spiderbeam are also available but are more suitable for home station, rather than portable, use. The following is an edited version of the detailed QST review. See also Chapter 4 for a discussion on making a 'homebrew' Spiderbeam.

So just what is a Spiderbeam? Forget aluminium elements and boom. Think full-size beam – a wire Yagi, with no traps, made of modern materials such as Kevlar and Velcro. Think quad antenna shape, but in a horizontal plane. The Spiderbeam website appropriately describes the antenna as "great for portable operations." The specs for front-to-back ratio call out what's typical of full size tribanders – confirmed by modelling and on the air tests. Yet the weight is 12 pounds and it breaks down easily – ready to haul around on trips.

The Spiderbeam that I built and reviewed is a tribander for 20, 15 and 10 metres. The beam features fibreglass poles put together in an X (the so-called spider), which hold interlaced wire Yagi antennas with three elements on 20m and 15m and four elements on 10m. The driven elements are separate almost straight wire dipoles interconnected by short transmission line sections to a common balun and fed with a single 50Ω coax cable. The director and reflector elements are constructed of wire bent into a 'V' shape.

You can choose other Spiderbeam options. The company also provides instructions to configure the tribander to cover 20, 17 and 15 metres for times like this when the sunspots are sparse! You can also choose a five-band version that adds 17 and 12 metres.

The Spiderbeam in all its glory.

The Spiderbeam folks intended their antenna to fit the market niche of hams who like to operate portable or want lightweight aerials. That meant simple and rugged. The designer came up with a method of tying the three driven elements of the monobanders to one 50Ω feed point through an included choke balun. Kevlar guy wires help keep the antenna in shape, as do the reflectors, directors and driven elements, strung with pieces of monofilament line between each element end and the X-shaped framework. An aluminium centrepiece that allows for several sizes of mast holds the fibreglass poles in the X framework. Much of the antenna is strapped together with provided Velcro strips.

A delivery van dropped off the 4 foot long Spiderbeam box. All of the parts fit tightly in the box. I spotted the large amount of Velcro first. Next I saw a formidable-looking large plastic bag chock-full of hardware. I dumped the contents of the hardware bag on the kitchen counter and sorted the mass of things into logical groupings; what a variety! I grabbed three plastic containers and put like groupings together. A nice surprise was that the company provided baker's dozen quantities. If the checklist called for 12 washers of a certain size, the company packed 14. The company knew hams could lose hardware when assembling and disassembling the antenna multiple times on portable outings.

The 26-page manual covers all construction options. This confused me until I spotted little boxes of text that said things like, "Kit contains pre-drilled enclosure" in the section about fabricating the enclosure. The pictures in the manual are pretty good, but close-ups would have helped even more.

The manual states that measurements must be made to within 1cm when cutting directors and reflectors. It also warns that using a metre stick over and over can introduce large errors. All dimensions in the manual are in metric units and I didn't have the suggested flexible plastic metric tape measure of at least 11 metres long in my toolbox. Converting every measurement in the instruction manual to inches beforehand allowed me to use my Stanley 50 foot steel ruler. For precise measuring and cutting, find a friend who can hold one end of the material to the tape measure.

The author of the manual wrote some directions with great skill. But other steps (things that weren't intuitive to me, it seemed) weren't clearly spelled out. I found myself rereading and asking other hams' opinion before proceeding. Some instructions may seem innocuous but turn out to be very important. My biggest error was cutting the monofilament stringers that would have insulators installed on the ends before attaching the elements. I cut per the dimensions in a chart. But what I didn't notice was the preceding paragraph that said the chart dimensions were the lengths between the insulators. After threading the monofilament through the insulators, tying the knots, attaching the elements and trying to attach the pieces to the spider, I realised my error – every length was cut too short! Time to get some 150 pound test saltwater fishing line and do them over.

The manual instructions said to tie insulators to the ends at the driven elements, measure the lengths and pull the wire through the insulator. To ensure the correct length, fold the wire back against the insulator and secure it with a cable tie. One wrinkle – the kit doesn't include ties, so I used electrical tape, which turned out to be easy to adjust for optimum SWR and has held up well so far.

The manufacturer supplied the black UV-resistant polyethylene insulators – 66 of them – for three purposes. Insulators go at the end of each wire element, are used as a guy tensioner for guy lines, and as centre insulators for the driven elements. The polyethylene pieces come with holes drilled in the ends, and slots reamed out in the centres. The manual suggested getting a pen with white ink, but where to find one? We located a silver permanent marking pen that worked great on the insulators, making it easy to sort the reflectors from the directors, and the 20 metre lengths from the 15 and 10 metre lengths. We marked everything we could, including the boom sections, so it would all go together properly the next time.

The toughest parts were building the driven elements and the balun. For me, fashioning the driven elements took the most diligence and time compared to any other part of building the Spiderbeam. The instruction manual warned once again of getting the accuracy to 1cm. Pieces of heat shrink tubing seal the connection of the cable lug soldered on to one end of each driven element half.

The manual warned also about ensuring that the wires of the transmission line sections, at the end of two driven elements and which are held with more shrink tubing, run exactly parallel and do not cross one another, or the feed line will have a 180° phase shift. The third warning is to use a hair dryer, not a lighter that can ruin the insulation, to shrink the tubing and to melt the glue inside it.

The 'web support' arms of the spider beam give an idea of its size. Each horizontal is about 33 feet across.

Putting together the balun took some care. The rugged Teflon coaxial cable didn't easily and tightly wrap around the ferrite toroid – in fact, the manual called for 12 turns, but I could get only 10 turns to fit. The tough inner jacket made getting at the braid a challenge. On the other hand, the Teflon coax cable is said to be rated to handle a full 2kW unlike other thin cable.

The picture showing how the braid and inner wire were to be fashioned for soldering was not large or clear enough for me, and the instructions didn't explain much more. I needed a friend to hold the toroid in the box while I soldered on the cable lugs.

If you picture a 33 foot by 33 foot cross structure, you're picturing the Spiderbeam accurately. The aluminium centrepiece is about 9 inches on a side, and the fibreglass poles measure about 1.4 inches in diameter. The photo above should give you an idea of the scale.

Out in the backyard the fibreglass poles go together quickly as O rings are slipped on between the appropriate sections and the decision is made as to which poles will form the boom and the lateral spreaders.

The manual suggests raising the whole assembly above the ground a few feet on a mast. Note that while a mast is not provided, there needs to be one dedicated to the antenna and assembled with it. There needs to be a total of about 3.5 feet exposed above the rotator or tower to allow mounting of the balun and attaching guy wires. We used a 5 foot, 1.5 inch diameter steel mast from our local amateur radio dealer and needed a second one to get clearance through the top section of our tower.

Next comes the positioning of the balun box on the mast with Velcro strips, and attaching the driven elements to the cable lugs. The manual again warns to ensure that the feed lines are not twisted, and that, indeed, the left feed point hardware connects to the left leg of each dipole.

Though the antenna weighs only 12 pounds, I couldn't have pulled the beam up the tower by myself. With a gin pole and a rope on a fibreglass spreader, the antenna with its many flexible guys and wires, insisted on catching on a gutter nail and a tree limb. Once the antenna was at the top of my gin pole, my reach just didn't allow me to hoist it on to the mast, Once again it really took two to make this project come together. The beam, installed up top and turning in the rotator (see photo) was a work of art!

The antenna in place as the author makes final adjustments to the coax loop to allow for rotation. The reason for the name is evident.

In order to verify the manufacturer's specifications, we ran multiple *EZNEC* passes. The results are summarised in **Table 6.2** and the SWR curves are shown in detail in **Fig 6.1** over the page. My impressions from QSOs tended to agree with what the plots showed.

So how did the Spiderbeam play in real life? I compared stations on it versus my dipole, and the beam brought nearly every signal up by 3 to 5 S-units. I found YO2IM, who was faint on the dipole, with its near Europe-facing main lobe, but copied him as a 559 signal with the beam.

I jumped into the Helvetia Contest with 100W from my Icom IC-701 on 20m, and the beam pointed toward Switzerland. I snagged three competitors and all were 2 to 5 S-units louder on the Spiderbeam. I didn't want to ask contesters to take time for an A / B test, so I moved to 15m. I called PY7JGO, who was 2 S-units louder on the beam over the dipole. He had a hard time hearing me on the dipole, but gave the beam a 599 report. Then I heard CX3EU very weak on the dipole, but 549 with the beam.

Another day I listened to European stations on 20m SSB. DJ7ZM was S6 with the beam pointed towards him, S3 off the back of the beam and on the dipole, and S1 off the side of the beam. Then I heard an S8 signal from

Manufacturer's Specifications	*EZNEC* Simulation Results
Forward gain (dBi) at 15m above ground: 20, 15, 10m: 11.7, 12.3, 12.6dBi	Mid-band; 11.6, 12.3, 12.7dBi
Front-to-back ratio across band: 20, 15, 10m: 15-20, 20-25, 20-25dB	10-22, 15-24, 15.6-24.6dB (28-29MHz)
SWR on 20, 15, 10m: <1.5, <1.5, <1.2	<1.4, <1.5, <2 (28-29MHz)
Length, width: 33ft. Weight: 12 pounds	

Table 6.2: Manufacturer's specifications and *EZNEC* simulation results for Spiderbeam lightweight triband Yagi.

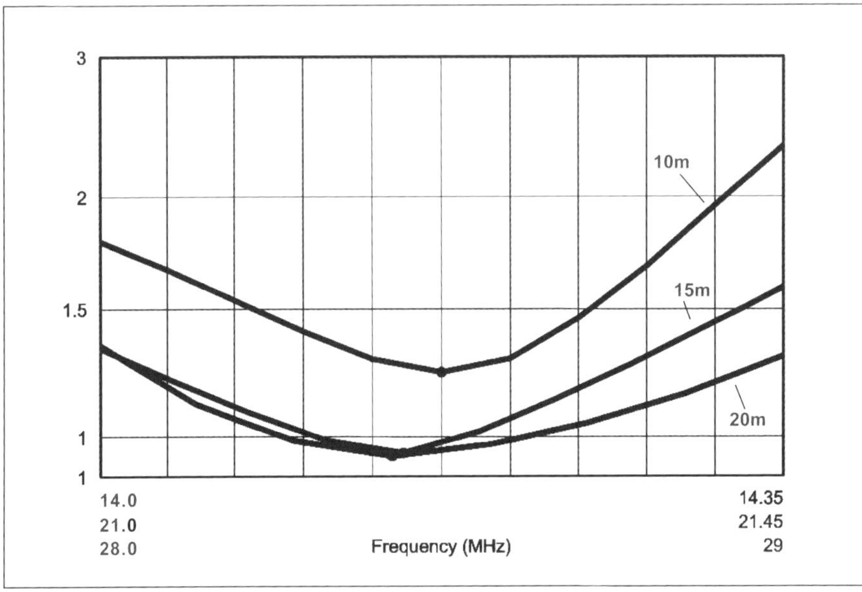

Fig 6.1: The SWR of the Spiderbeam as predicted by *EZNEC*. Our measured values were similar, although we shifted ours downwards to optimise for CW operation.

Florida who was S2 on the dipole, S5 off the back of the beam and S3 off the side.

With an extra director and more gain on 10m, that band should be even better than the others for times when the band opens. Unfortunately, that didn't happen during the days I tested the beam.

To summarise, the Spiderbeam is an interesting innovation and we applaud Con Paul, DF4SA, for making the construction details available to all at no cost! If you decide to build the Spiderbeam, from the kit or on your own, you will measure and cut a lot of things. If you have a little trepidation as I did, you'll triple check every piece before cutting. That means the time it takes to build this beam can go way beyond what the manual says. Even with help from friends, preparation and installation took about 60 hours. If I were to build another Spiderbeam, however, I believe it would take somewhat less time, having learned the knack for some of the tasks. After the first fabrication and assembly, with everything precut and marked, it should only take a few hours to pull it from a travel container and have it ready for use.

If you want a lightweight low-loss multiband Yagi without traps this beam is for you. If you enjoy operating portable, doing fun DXpeditions, prefer having one feed line or you're ready to go beyond just dipoles and verticals, consider the Spiderbeam, especially if you have a friend who can help with assembly.

When I first thought about the Spiderbeam, I was sceptical about how it would hold up in bad weather. Days before finishing this review, a fast, furious rainstorm hurtled through my town packing 40mph winds for 35 minutes. The beam turned a bit to and fro' with the wind currents as I watched out the deck door, but nothing broke. My scepticism was swept away with the remnants of the storm. Maybe lightweight materials fare better than heavy stuff.

The lightweight 10-15-20m Spiderbeam kit costs 349 euros including 19% VAT (or 293.28 euros for export outside the EU), excluding shipping, from the spiderbeam.com website. In North America it costs US $439 excluding shipping from spiderbeam.us

ULTRA LIGHTWEIGHT PORTABLE HEXBEAM

Steve Telenius-Lowe, PJ4DX (February 2020)

The UK manufacturer of the G3TXQ-designed Hexbeam antennas, Ant David, MW0JZE, visited Bonaire in 2017 and brought with him one of his antennas. Ant makes two versions of the Hexbeam: one designed for home station use, and a portable version: the 'Ultra Lightweight D-HEXpedition Portable Hexbeam' to give it its full title. It was this lightweight version that Ant brought to Bonaire and I took advantage of using it at my station.

Weighing in at under 6kg and packed into a bag only 1m long, the MW0JZE lightweight Hexbeam certainly qualifies as a portable antenna. Unlike some other manufacturers' Hexbeams, and the Spiderbeam reviewed earlier, this Hexbeam is *not* a kit. Obviously it still requires assembling, but all the elements come pre-assembled and it is just a matter of putting together the fibreglass frame and threading the wire elements through the insulators which are pre-fitted on to the frame. No measuring is required and no tools other than an Allen key, which is supplied.

The Ultra Lightweight D-HEXpedition Portable Hexbeam used on the 4W6A IOTA DXpedition to Atauro Island in Timor-Leste, September 2011.

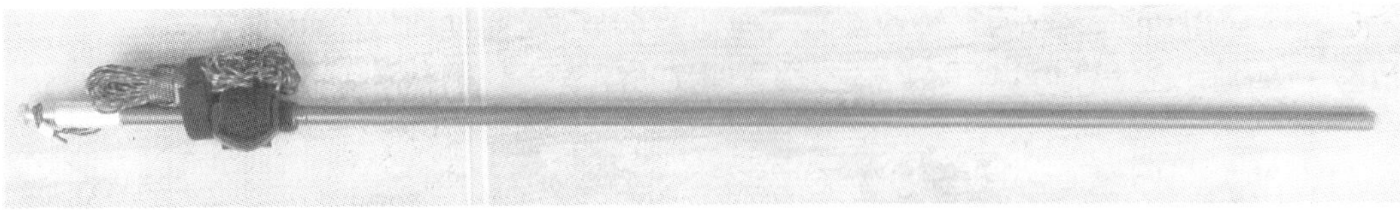

Laura, MW6INK, assembling the frame. The colour-coded spreaders are slotted into the hub (far right of photo) and into each other.

There is no electrical difference between the 'standard' Hexbeam and the portable version; the only difference is in the size and weight of the materials. This antenna covers six bands: 6, 10, 12, 15, 17 and 20m (some other manufacturers' Hexbeams cover only 10 to 20m). There are two elements on each band, a driven element and a reflector.

Before the elements can be connected the frame must be assembled. At the centre of the antenna is a hub into which colour-coded fibreglass spreaders (above) are slotted. The spreaders are very like those used in dome tents except they have pre-measured cords and insulators attached at the appropriate places. The cords have two purposes: to keep the six 'arms' fixed at 60° to one another and to create the 'upside-down umbrella' shape which in turn helps to provide the correct spacing between the elements on each band.

Six separate packages contain the wire elements. Each band's elements come as a continuous length: a half driven element separated from the reflector by an insulator and a precisely-measured length of spacing cord, then another insulator and spacing cord and finally the other half of the driven element.

The rigid balanced feed system to which the driven elements are attached.

Taking each band one at a time, the easiest way to add the elements to the frame is to start at the back of the antenna and thread the end of one of the half driven elements through the insulators towards the feed point above the central hub. Then do the same thing with the other driven element half. It is quicker to do this with two people, one taking the left side and the other the right. Start with the 6m-band elements and work your way up to the 20m elements. Each lower-frequency band is mounted somewhat higher up the frame and the 20m wires end up about 1m higher than the 6m elements.

The completed Ultra Lightweight D-HEXpedition Portable Hexbeam is light enough to be carried by one person. The insulators and spacing cords between the driven element halves and the reflectors can be seen clearly.

The feed points are on a rigid balanced feed system consisting of two lengths of square-section aluminium which together provide a 50Ω impedance. The two half driven element ends are connected to their respective feed points by wing nuts which are simply tightened by hand: once again, no tools are required. Your 50Ω coax feeder is connected at the top of the rigid feed system.

No balun is provided as standard – and the Hexbeam appeared to work perfectly well without one – although a 1:1 balun is available from Hexbeam UK as an optional extra.

You end up with an antenna that measures 6.5m x 6.5m and which is light enough to be held up by one person. It can be put up on a lightweight push-up mast and rotated easily by hand – we attached strings in two places on the frame in order to turn the antenna. Although it may look rather frail it offers little wind resistance and so is actually quite stable in the wind. 'Storm cords' are available as an option for particularly windy locations and would probably be advisable for peace of mind, although these were unnecessary here on Bonaire. The first time I assembled the Hexbeam by myself it took me just over an hour to build it from scratch, though I am sure it would be quicker on subsequent occasions now that I know what I'm doing!

Band (metres)	Forward gain (dBd*)	Front-to-back (dB)
20	3.8	22
17	3.2	19
15	3.5	16
12	3.0	13
10	3.6	16
6	2.7	11

Table 6.3: Claimed gain and front-to-back figures of the G3TXQ Hexbeam. Both the standard Hexbeam and the Ultra Lightweight portable version described here have the same claimed figures. *Add 2.15dB to these figures for dBi.

One of the great advantages of the G3TXQ Hexbeam design is its wide bandwidth. What this means in practice is that there is no need to adjust any of the element lengths and the SWR is low across the band. For example on 20m, at my location the SWR was 1.0:1 at around 14150kHz and below 1.4:1 across the whole band. The same was true on 17, 15 and 12m. On 10m, the SWR only rose above 1.5:1 at about 29430kHz and it was around 2:1 at the top of the band. Clearly proximity to objects such as houses or trees, the antenna's height above ground, the length of feeder and amount of feeder loss will all have an effect on the SWR figures but, whatever the absolute values, the design means the bandwidth will be good and the SWR low.

The claimed gain and front-to-back ratio figures are shown in **Table 6.3**. After doing numerous 'A/B' checks with a commercial 5-band Spiderbeam at about the same height above ground, the figures in the table seem overly modest to me. I found no discernible difference between the Hexbeam and the Spiderbeam on 12m and 17m, on which bands both antennas have two elements. There was also little difference on 15 and 20m, where the Spiderbeam is a 3-element Yagi, with some signals appearing to be the same on both antennas, while others were a few dB (less than an S-point) down on the Hexbeam compared with the Spiderbeam. It was really only on 10m, where the Spiderbeam has four elements, that there was a noticeable difference between the two antennas. Unfortunately 10m was almost 'dead' at the time of the tests but on the few signals found the Hexbeam was about one S-point down on the bigger Spiderbeam (no tests were carried out on 6m).

To conclude, the lightweight Hexbeam is ideal for portable operations, National Field Day, DXpeditions etc where some gain over dipoles, G5RVs or verticals is required. It would be particularly useful for a single-operator activity, where a larger or heavier beam may be impossible for one person to handle without assistance.

The Ultra Lightweight D-HEXpedition Portable Hexbeam is available direct from the manufacturer, Ant David, MW0JZE, trading as 'Hexbeam UK', via the website at g3txq-hexbeam.com It costs £460 including VAT. Optional extras include a 1:1 balun, 'storm cords' for use in windy conditions, and a fabric carrying bag.

THE FOLDING ANTENNAS HEXBEAM

Reviewed by David Aslin, G3WGN
RadCom, April 2015

The Folding Antennas Hexbeam is another contender in the 'portable HF beams' category. Designed and produced by Christian Sack, DL1LEU, electrically the Folding Antennas Hexbeam is the same as a G3TXQ Broadband Hexbeam, but its innovative mechanical design with its ability to be folded into a manageable package leads to its portability. While G3WGN reviewed the Folding Antennas Hexbeam at his home station, he pointed out that "it's an ideal antenna for SOTA, NFD, IOTA, DXpeditions and special events etc."

For Hexbeams, there are two principal mechanical design types – planar (all element spreaders are horizontal) or 'upside down umbrella' that requires the spreader arms to be tensioned into an upward curve. DL1LEU chose the planar version. Six fibreglass spreaders with folding joints hold an M-shaped driven element (dipole) and a U-shaped reflector for each band in an overall hexagonal shape. There is a single feedpoint with an SO-239 connector.

Christian's particular forte is in precision moulding of plastics. His skill in this area is evident in the quality of the Folding Antenna components – the fit and function of the wire clips, element end caps and the coaxial centre tube are all first class. There are excellent pictures of the components on the Folding Antennas website. As standard, all HF bands from 20 – 10m are included. Gain is claimed as 3 – 4dB over a dipole and front-to-back 20dB (varying by band). SWR is specified at less than 2:1 across all the HF bands except 10m.

The Folding Antennas Hexbeam in the air.

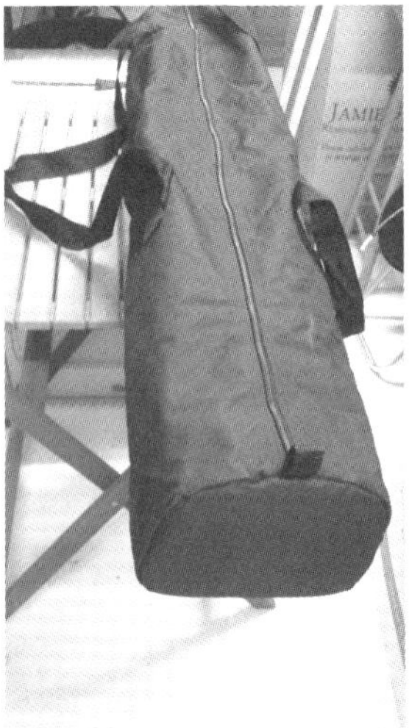
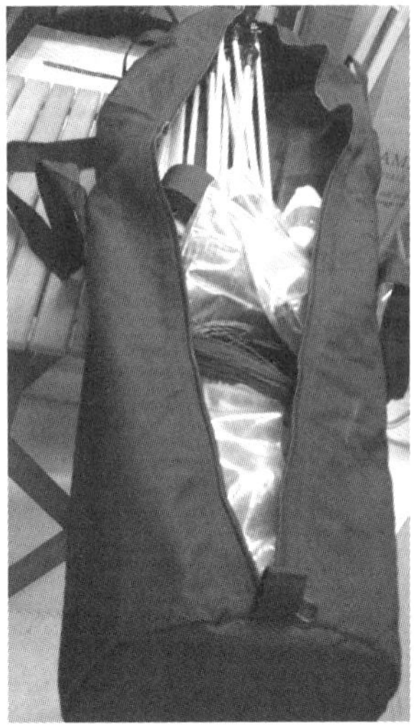

The Folding Antenna packed in the optional carrying bag.

Options that came with the test antenna were:

- The carrying bag. In my view this, or something similar, is essential if you are going to use the antenna away from home, to avoid snagging and tangling of the elements during transport when the antenna is in its folded format;
- 'Storm brackets'. These are recommended by Christian for permanent (or longer term temporary) installations. They prevent the lifting of the spreader arms under windy conditions. The brackets were fitted to the test antenna;
- A large 31-mix ferrite core for constructing a common mode feedpoint choke. I wound five turns of Aircell 7 feeder around the core and used cable ties to secure the resulting choke to two of the spreader arms. The choke follows K9YC's recommendations **[1]**;
- The review antenna was fitted with the optional elements for 6m.

The Folding Antenna arrives well-packaged in a box, containing the antenna parts inside their carrying bag. It is best described as a kit – there is considerable work to do to prepare all the parts. The clips that hold the wires to the spreaders and the clamps that connect spacer cords to the end of the wire elements arrive as parts on a set of plastic 'trees' – to be cut off with a sharp knife and fitted together just like in the Airfix kits I put together as a lad.

I assembled the antenna on a slow-and-steady basis so I could document every step with notes and photographs. I spent a total of 21 hours spread over several weeks putting it together and adjusting all six bands to optimise the tuning. Others report doing the job in 8 to 10 hours, which is certainly possible but I would caution prospective builders to take care to follow

The Folding Antennas Hexbeam arrives as a kit – with a lot of parts!

David, G3WGN, starting construction.

The completed antenna, ready to go.

carefully and exactly the comprehensive written instructions that Christian provides. If I missed even a small note, I paid for it by having to redo part of the work!

Ideally the Folding Antenna should be mounted at a minimum of half a wavelength above ground at its lowest frequency of operation – i.e. about 10m / 33ft in this case. I have a push-up telescopic fibreglass mast from Moonraker **[2]** that will go to this height with a dipole at the top, but will probably be overstressed with the 7kg weight of the Folding Antenna at full height. The antenna was elevated to 18 – 24 feet for performance testing, by nesting the top sections of the mast. I have recently taken delivery of an aluminium 11.3m rotatable mast from VPA Systems **[3]**. With a Yaesu rotator at the base of this mast it is now possible to achieve full rotation of the Folding Antenna at elevations up to 10m / 33ft.

It is not a good idea to attempt erecting this antenna from the horizontal. Even with the optional 'storm brackets' the antenna is unwieldy and may end up with some of the cords and / or elements entangled. A push-up telescopic mast is the ideal support, whether fibreglass or metal.

After assembly, it was time to test the SWR on each band. The test conditions were:

- Antenna height above ground: 7m / 23ft;
- Weather: dry (see later for why this is an important test parameter);
- Feeder: 10m of Aircell 7 coax from feedpoint to ground level (including the five turns around the ferrite core near the feedpoint);
- SWR measurement: RigExpert AA-230PRO connected at the end of the Aircell 7 feeder. Full scan of each amateur band to display the SWR curve.

Without adjustment and constructing carefully the elements to the dimensions in the assembly instructions, several of the bands already showed an SWR that measured less than 2:1 at the band edges and an SWR minimum at the top end of the CW sections of the band – which is where I wanted the resonance to be.

The end insulators provided with the antenna allow adjustment of element lengths by a simple sliding release knob and pulling the wire through to make it shorter or longer as needed; then pushing the knob back into place to retain the wire. This is slightly fiddly to do, but overall a satisfactory way to bring the tuning of the antenna to the desired point in each of the bands. Having achieved the desired tuning, there is a need to readjust the position of the wire elements on the spreaders, and maybe adjustment of the spreader cords will be required. This takes a methodical, symmetrical adjustment process and while a little tedious, is not difficult. The end result is an antenna properly tuned to each band, with the spreaders as straight as possible and horizontal.

The 6m band showed a very broad resonance, which made adjustment more difficult. This requires further work on my part to optimise the tuning.

One of the known characteristics of this type of antenna (in common with all Hexbeams and most wire Moxon designs) is that accumulation of rain on the elements causes detuning. The dip in the SWR curve shifts frequency when it's raining and the effect remains while the antenna is wet. This is nothing an ATU can't cope with, but if you are using an amplifier or radio without an ATU, you may have to retune between wet and dry. **Table 6.4** shows the magnitude of the effect with the antenna at 7m / 23ft. Note that the resonance shifts outside the band if the Folding Antenna has been tuned for the CW end of the band. Some judicious juggling of the tuning should allow reasonable coverage of the bands in both wet and dry. In practice I've found no noticeable difference in performance between 'wet' and 'dry', in spite of the shift of the resonant point.

Patience in assembling, testing and adjusting the antenna was amply rewarded. Over a period of three days of operating immediately following completion of the antenna, I broke pile-ups for SU9VB, S21ZBC, 3DA0ET, FR/F5UOW and went on to work XZ1J, Z81X, W8A (American Samoa) and FT5ZM during the first couple of weeks of operation. All were worked running 100 or 200W with a 'barefoot' Yaesu FTdx5000. More recently, using a K3 with 100W, record-breaking scores in low-power single operator categories in the 2014 CQWW CW (15m single band) and ARRL 10m contests have been the highlights.

Band	SWR Curve DRY		SWR Curve WET		Δf kHz
	Min MHz	Min SWR	Min MHz	Min SWR	
20m	14.087	1.12	14.041	1.08	–46
17m	18.104	1.13	18.010 *	1.12	–94
15m	21.075	1.38	20.850 *	1.78	–225
12m	24.977	1.29	24.775 *	1.19	–202
10m	28.093	1.02	27.880 *	1.01	–213
6m	51.154	1.47	51.278	1.93	+124

Table 6.4: Frequency shift of resonant point between dry and wet conditions. (* Wet elements shifted resonant frequency out of band limits.)

Lists of exotic DX and contest results are not a precise indication of antenna performance, but few of us have access to an HF antenna range, so this list will have to do as a performance indicator. I also used the Reverse Beacon Network to check front-to-back performance and found typically 12 – 20dB depending on band, distance and conditions; this is consistent with what others have found for the G3TXQ design. Further experience in both contests and casual DXing have brought a very satisfactory set of contacts and much enjoyment – well beyond what I expected from a small antenna at the low height of 7 – 8m. I even got a genuine 599 report from my friend Tom, W7RN, in Nevada, when the Folding Antenna was down at 2.3m / 7.5ft in height because of the high winds! That might have been something to do with Tom's setup – a stack of four 6-element long boom Yagis on a 140ft tower for 10m – but, hey, if I get DX QSOs from my postage stamp sized garden, I'm happy. In January 2015 that was topped by working EP6T and C5X through huge pile-ups, also with the antenna at its retracted 2.3m height. Just to be clear, I'm not recommending operating the antenna at such a low height. I am lucky enough to have an exceptional location – at 220m / 720ft with 360 degrees of clear take-off and downslope in all key directions – which not everyone will have access to. The antenna does work at low height, but it will work far better if you put at around 9m / 30ft or higher.

The severe winter winds and rain of winter 2013 – 14 and 2014 – 15 were a test for any antenna. To avoid damage the support mast was often telescoped down to minimum height – about 2.3m in my case. However, the antenna was in the air at around 7m / 24ft during some quite severe gusts of 35 – 50mph. It showed no signs of distress under these conditions, although its gyrations and waving about were alarming to watch. Now, after 24 months in the air, it still looks like new.

Does it really fold? There are a number of attempts to document the folding process on the web; the most useful I found is from Lars, PH0NO **[4]**. I followed the steps he recommends and had no trouble with folding and

The partly folded antenna.

The completed antenna under test.

unfolding (having naturally got into quite a tangle with my first unguided attempt!)

Overall, the antenna gives a very good account of itself on the air, in the air, and folded on the ground. There are a couple of items that are opportunities for improvement:

● I could get the antenna to tune where I wanted it on all bands 20m to 10m, but some of the antenna clips had a tendency to slip, causing detuning of the related band. Christian recommends using superglue to fix the clips, but I'd want to be very sure that I had the antenna tuned exactly how I wanted it before using that rather permanent method of fixing the clips. I'm experimenting with Stauff clamps which have so far worked well during my experiments with and evaluation of the antenna. They are attached beside the Folding Antenna clips to prevent slippage. Once the experiments are over, then I'll use the superglue.

● I found 6m quite difficult to tune. The dip in the SWR curve is quite broad and shallow. This requires more investigation. I have not used the antenna on 6m at the time of writing.

● The element lengths in the assembly instructions leave a short (10 – 15cm / 4 – 6in) 'tail' of wire hanging at the end of the insulators. This allows for adjustment and can be trimmed during tuning. On some bands, 5 – 10cm of 'tail' remains after adjustment. Anyone familiar with making wire dipoles, for example, would instinctively wrap the tail back on to the element itself. That's what I did anyhow. This had the effect of significantly detuning the band in question. It seems from conversations with other builders and with Christian, that the tails are intended to be left free. This is not clear in the assembly instructions.

These issues do not detract from a very worthwhile, compact, lightweight and effective transportable antenna suitable for fixed and portable operation. It is not difficult to build and set up and so far has survived all that our island weather systems could throw at it. The support from Christian, DL1LEU, of Folding Antennas is first class.

The 5-band Folding Beam kit costs 425 euros for outside the EU, or 506 euros inc 19% VAT within the EU. Shipping is extra and options include carrying bag, 6m elements, storm brackets, ferrite core for choke etc. The antenna can be ordered from www.foldingantennas.com/gb.html

STEPPIR CRANKIR PORTABLE HF–VHF MONOPOLE

Reviewed by Joel R Hallas, W1ZR
QST, June 2014

The CrankIR is a very versatile portable HF / VHF antenna system that can serve many functions effectively without the need for additional supports or gear. In recent years it has become a popular choice for use on DXpeditions, particularly those located adjacent to the sea which can take advantage of the additional low-angle gain provided by a vertical.

The CrankIR is designed to be a portable HF / VHF vertical monopole that can be employed without the need for any on-site supports, such as trees, buildings, or poles. This can make for a very independent and flexible portable operation. In order to make all of that come together, you must carefully select from the extensive option menu to make sure you have all the pieces for your preferences.

The CrankIR is a departure from SteppIR's remotely adjustable Yagi antenna designs in that all adjustments are made at the antenna using hand cranks. This antenna is designed to be stored and carried in a 24in (61cm) bag. When unfolded and erected, it can be adjusted to serve as a monopole for any frequency (one band at a time) between 40m and 2m, or extended to 80m with an optional kit [*see note at end of review – Ed*].

The basic antenna package does not include a base, so some thought must be given to make sure that it can be properly secured. While SteppIR offers a choice of two compact collapsible bases, the antenna could be clamped to a railing or fence post. As with any antenna, make sure that you place the CrankIR away from any power lines and in a spot where people won't come in contact with it.

The antenna support consists of telescoping fibreglass tubing that can extend to 16 feet (4.9m) above the feedpoint. This length can be used as a typical monopole support for the higher frequencies. The innovative part of this design is that for lower frequencies, the wire can be cranked further to take a turn over a cross T at the top of the support and then can run back towards a horizontal strut behind the crank in a kind of folded configuration, reminiscent of the 'trombone' elements in some of the SteppIR Yagis.

The monopole element consists of a single wire that is pulled out from a reel that is part of a dual reel with

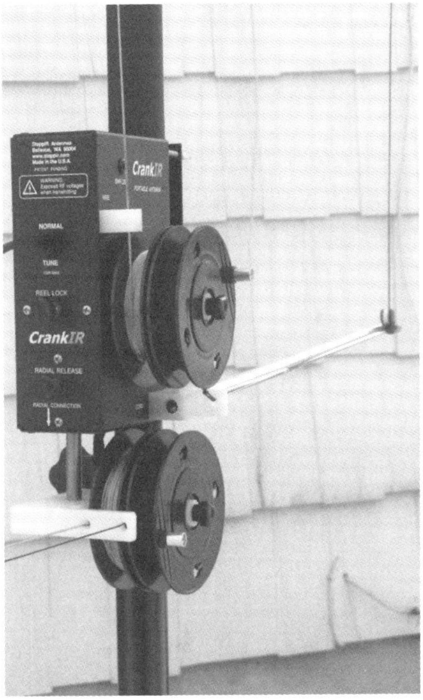

Close-up of the crank and instrumentation assembly.

hand crank. The second reel takes up the twine as it comes down from the top. The T tensioner section at the top of the mast maintains separation between the up and down paths and automatically adjusts the tension of the system to account for the changing diameter of the reels as the wire and twine is paid out and pulled in.

As with all monopoles, a radial or counterpoise is required. The basic antenna is provided with a lug that can be used to attach user-supplied radial wires, or an optional radial reel can be employed to provide an adjustable radial. It is recommended that elevated and insulated radial(s) be employed for maximum performance.

The photo to the right shows the details of the crank and instrumentation assembly with the monopole and radial crank set up for operation. The monopole wire leaves the upper reel and heads skywards with the far end secured by the twine, and returns via the strut on the right. The lower radial wire and twine are going off to the left from the lower crank. The centre finger nut can lock the reels together so that tension is maintained while the length is adjusted. For deployment or disassembly, the finger nut is loosened to allow the two reels to operate independently.

Very much to SteppIR's credit, the entire antenna and the base we selected can be assembled without tools. All the fasteners are engaged with captive wing nuts that stay in place. As part of the package, SteppIR provided a comprehensive 48-page full-colour assembly and operation manual. Follow the instructions exactly to avoid problems in making it all come together.

After the wire is run through the guides, with the reels uncoupled using the CORD REEL LOCK / DRAG thumb nut, the telescoping mast can be raised as much as required for the bands to be used, pulling the wire with it. Each section has a friction fit on to the next that will hold it in place, although the instructions suggest that tape be used to secure the joints. I second that suggestion – I found it quite dramatic to have the whole mast collapse above my support hand. No harm done, but I quickly went for the tape.

This CrankIR is clearly designed for operation from a nearby picnic table or other portable arrangement. The rough monopole antenna wire length is indicated by a colour coded indicator on the wire, one for each band. For final tuning, a built-in SWR indicator with LED display can be used to adjust the antenna and radial(s) while driven by the connected transceiver. While that can work fine in nice weather, Connecticut's below-freezing temperatures during the evaluation were not compatible with picnic table operation, at least not by me. For operational testing, I used my trusty RigExpert antenna analyser to set the frequency and then used a transceiver in my shack – much cozier. If you do use the built-in SWR indicator and your transmitter to adjust the antenna, keep safety in mind. Use very low power and avoid touching the antenna conductors with RF applied.

The optional ground radial kit works almost the same way, except instead of the T tensioner on top of the mast, there are small pulleys attached to lengths of elastic cord that can be attached to two different points to provide the needed separation if the folded configuration is used. The support base we had could be extended upwards to elevate the base as desired. This will reduce the main elevation lobe and also allow the radial(s) to droop to improve the match, if desired.

Note that while the typical configuration is to have the monopole and radial length similar, it is not necessary. If the monopole is longer than the radial, for example, it will act much like an off-centre-fed monopole, providing somewhat improved performance on the bands that don't need to be folded, along with a higher input impedance. In a similar manner, the additional 6 feet (1.8m) of support provided with the 80m kit will not get to a quarter-wave monopole, but resonance can be achieved by extending the radial a corresponding amount.

While Connecticut's late winter weather was not optimum for this kind of endeavour, I did have some opportunities to run the antenna through its paces, starting on 20m. Once the antenna was set up, both sets of reels were recoupled by tightening the REEL LOCK/DRAG nut. Then both the monopole and radial cranks smoothly adjusted element lengths with a precision and ease that I'm not accustomed to for wire antennas. I used the colour-coded tags to set the monopole length and adjusted the radial for the best match.

An *EZNEC* **[5]** model predicted that the single radial would result in an antenna with a strong preference for transmission in a wide main lobe in the direction of the radial, so I pointed it towards Europe and compared the results to a 30-foot (9.1m) high three-element tribander. As expected, the Yagi had a 1 to 2 S-unit advantage for US stations at higher angles. Performance towards Eastern Europe was much closer, but still with some advantage to the Yagi.

I next set up the CrankIR in full folded configuration for 40m. Here, with the strut and radial both pointed towards Europe, the predicted pattern was even more dramatic with a deep null to the rear, especially at higher angles. I compared the CrankIR to my bidirectional 50-foot (15.2m) high 135-foot (41m) horizontal centre-fed two half-waves in phase antenna during the SSB weekend of the ARRL International DX Contest. While the horizontal antenna was pointed both towards Europe and California, the CrankIR was just pointed towards Europe. The results were quite dramatic, especially when viewed on my LP-Pan panadapter. With the horizontal antenna, the

European stations were small blips appearing in gaps in the solid wall of US stations. Switching to the CrankIR dropped the 'wall' dramatically while the DX stations, especially those at greater distances, were fairly close to the levels with the horizontal. It meant that I could remove the attenuator from the receive path and have good copy and contacts.

Because I'm on the East Coast, for ARRL Field Day I would simply install the antenna in the other direction. For those planning for Field Day in Kansas, however, the use of two radials 180° apart, perpendicular to the strut and T tensioner, will give a pattern within about 0.8dB of being omnidirectional, with about 1dB of additional intensity in its favoured direction (based on *EZNEC* results that assume uniform ground beneath the radials).

The latest (Mk IV) version of the SteppIR CrankIR, covering 40m – 6m, is available in the UK from Nevada (www.nevadaradio.co.uk) for £358.29. The 80 – 6m version costs £583.29. A 6ft tripod stand is an optional extra and costs £66.63.

BRAVO-7K PORTABLE VERTICAL DIPOLE

Reviewed by H Ward Silver, N0AX
QST, March 2012

The Bravo-7K set up on the north coast of Puerto Rico.

Ward Silver, N0AX, wrote that his planning a summer beach holiday (with radio) created the need for a portable antenna. His requirements were:

- *40 to 10m, including 17 and 12m;*
- *High efficiency without radial wires;*
- *Light weight;*
- *One-person assembly and adjustment;*
- *Pieces sized 4ft or smaller when disassembled;*
- *No control boxes or power requirements.*

Ward wrote "I found lots of antennas that met some of the requirements, even most, but I was beginning to think compromise was going to be necessary. I recalled that Tom Schiller, N6BT, along with the continuing expertise of Team Vertical, was returning to the antenna business and discovered to my delight that he was just finishing the Bravo-7K. It appeared to satisfy every one of my needs!"

The Bravo-7K is 18 feet (5.5m) tall and a bit over 18 feet wide when all of the elements are fully extended for 40m operation.

The elements are made of telescoping aluminium tubing held in place with stainless steel hose clamps. The total weight is 13 pounds (5.9kg) and the longest piece is 36in (91cm) long. The antenna has a single vertical element and two 'radial' sections. However, the antenna is not a typical quarter-wave ground-plane design, and the 'radials' aren't ground-plane radials in the usual sense. An evolution of N6BT's previous vertical dipoles, it is an off-centre-fed configuration with the lower half split into two radials.

Loading coils at the feed point are inserted into both elements and the total amount of inductance is changed for each band. Even with all loading coils inserted, the claimed efficiency exceeds 87% (0.59dB loss) on 40m [Update December 2019: according to the manufacturer's website, the antenna is 92% efficient – *Ed*].

Loading inductors are added in series with the vertical and radial elements on different bands. No inductance is required on 10 and 12m.

Tuning the antenna for the different bands is done manually by changing the element lengths and moving a shorting jumper on the tuning coils. The photograph to the left shows the inside of the coil enclosure. The two coils on the right are inserted between the feed point and vertical element on 40 and 30m. The four coils at the left are inserted between the feed point and the radials on 40 to 15m: no coils are used on 12 and 10m. Changing the element lengths entails loosening the hose clamps and adjusting the telescoping tubing lengths. The available remote-tuned version (Bravo-7KR) switches the coils in and out with the element lengths unchanged.

My first impression was that the antenna is simple and sturdy. The supporting tripod is welded. Mounting and assembly hardware is stainless steel and big enough to do the job. The coil enclosure of thick plastic comes with an O-ring if you want to seal the box, although that's not necessary. The antenna is rated for use with legal limit power [1500W in the USA – *Ed*] and it appears to be built to handle the voltages and currents involved. (It was tested with a maximum of 100W.)

Assembling the antenna is straightforward and the instructions are clear. Following the element section length table and setting the jumpers requires very little head-scratching and no 'tweaking' is required once the antenna is configured for a particular band.

The antenna is specified for a minimum SWR of 1.3:1 at resonance on all bands and I found that to be true except on 40m where resonance was below the band. N6BT reports that is not uncommon because of different ground types or mounting farther away from ground and includes an alternate set of 40m coils with the antenna. Changing the coils resulted in the expected performance on 40m after stretching the 40m coil a small amount to obtain minimum SWR. The only band that required tweaking to get specified performance was 40m. I also found the antenna performed well on 20 to 10m while set for 17m and using an auto-tuner to take care of the resulting impedance mismatch.

Band changes take a couple of minutes. Starting with the radials, adjust each section to the proper length. Change the jumper position by loosening a nut and moving the terminal. Tip the antenna over and adjust each vertical section for the proper length. Return the antenna to vertical and you are back on the air.

On the air, the Bravo 7-K plays like a full-size ground-plane antenna with a decent field of radials. It plays well at the beach and in the back yard – I even put it on the roof for some evaluation. One of these days I would very much like to perform a 'shootout' comparison to a full-size quarter-wave monopole.

As with all antennas there are strong and weak points. The strongest point is performance similar to what I'd expect from a full-size antenna but in a reduced-size package. The second strong point is the extreme ease of transportation and set-up. I am completely unpacked and on the air within 30 minutes of arriving at a portable location. My entire station, including the antenna and feed line, can fit comfortably within a hard-shell golf bag and still meet a weight limit of 50 pounds (23kg). (I recommend carrying radios by hand, though, if you are flying.) This makes it quite easy to 'grab and go' for a quick afternoon of operating.

What I don't like: although sturdy, the antenna will tip over in a strong wind if not tethered. A tent peg and a short length of guy rope holding from the wind's direction are sufficient to solve the problem.

Measuring section lengths gets old, so I used an engraving tool to mark lengths and band names on each section. The feed point requires a 'pigtail', so I made a short adapter section for attachment to a PL-259 coax connector.

Next to the tripod in the lead photo you can make out a coiled-coax choke balun wound on a plastic water bottle – this helps isolate the feed line's outer shield surface from the antenna's radiated field. I also turned the coil mounting screws around to extend outside the coil enclosure and replaced the standard hex nuts with stainless steel wing nuts – this speeds up the band change process. In fact, I can now assemble the antenna using only a miniature pocket multi-tool.

I have used mobile whips, ground-planes, sloping dipoles and inverted-V antennas in portable operation and this one is by far the best I've encountered to date. It's in the portable station kit to stay!

As of January 2020 Tom Schiller, N6BT, trades under the company name Next Generation Antennas. The Bravo-7K can be ordered from the www.nextgenerationantennas.com website. It costs $349.

N6BT Q-52 PORTABLE HF YAGI

Reviewed by H Ward Silver, N0AX
QST, September 2012

The Q-52 is a compact rotatable Yagi antenna with gain that covers the upper five HF bands. Weighing in at only 18 pounds (8.2kg) it may be the ideal choice for a portable operation when gain over a dipole or vertical is required.

The Q-52 is a five band, two-element Yagi, hence the model name (there is also a Q-51 rotatable dipole). It is full size on 10m and loading inductors are switched in with relays to lower the resonant frequencies of both the driven and reflector elements for use on 12, 15, 17 and 20m. A control box turns the relays on and off from the operating position through a multi-conductor control cable. Band switching is instantaneous and could probably be adapted to automated switching schemes.

A separate switch on the control box opens the reflector element for 'bidirectional mode', using the driven element as a rotatable dipole. A small inductor is used for a hairpin match to 50Ω.

Ferrite beads create a choke balun between the unbalanced coaxial connection and the symmetrical loading networks. Thus, no additional balun is needed on the feed line. The control cable, I discovered, *does* pick up RF and *does* need a coiled or ferrite bead choke.

The Q-52 Yagi installed on a rooftop mast about 30 feet (9.1m) above the ground. The antenna mounts easily on a 1.6in mast and can be turned with a light-duty rotator such as an AR-40.

The Q-52 control box has a single rotary switch for band selection and a toggle switch to open the reflector element for bi-directional mode.

The driven element's relay and inductor assembly with the control cables to the operating position and to the reflector element attached to the terminal strip.

Physically, the antenna is light in weight (18lb / 8.2kg) and has a turning radius of only 9 feet (2.74m). The boom is just less than 9 feet long and a light duty rotator such as the Hy-Gain AR-40 would suffice. For portable or Field Day use, it could be supported by a guyed push-up mast or fibreglass pole up to about 20 feet (6m) and rotated by hand.

A plastic enclosure mounted on the boom houses a circuit board, loading inductors, switching relays and a terminal strip as shown in the photo above right.

All supplied tubing is drawn aluminium and the hardware is stainless steel. The plastic enclosures are high impact ABS with sealing O rings.

Table 6.5 summarises the published specifications. At 30 feet (9.1m) in height, forward gain is specified to be a minimum of 3.3dBd on 20m and increasing with frequency. Front-to-back ratio is given as typically 8 to 10dB. Each antenna is assembled and tuned at a height of 30 feet before shipping.

As with other products I've purchased from N6BT over the years, the mechanical portion of the antenna is robust and well built. Holes that were supposed to line up did so. U-bolts and other hardware were high quality. Welds had no gaps or blobs.

The photo over the page shows the parts ready for as-

Boom length: 9 feet (2.74m)
Turning radius: 9 feet.
Weight: 18 pounds (8.2kg)
Wind survival: 100 mph
Wind load: 2 square feet
Feed point impedance: 50 ohms
Power rating: 600W SSB
Minimum gain: 3.3dBd (20m, 30ft / 9.1m height)
Typical F/B: 8 –10dB (30ft / 9.1m height)
Power requirement: 12V DC
Price: $689

Table 6.5: Manufacturer's specifications for the Q-52 portable Yagi (amended with updated information from manufacturer's website, December 2019).

sembly right out of the box. The electrical relay assemblies appeared to be soldered well and the coils were made from heavy duty wire. Following the instructions is important but not difficult. I would like to see the photos and drawings integrated with the assembly checklist, though. Total assembly time was a little over an hour from opening the box.

One thing I definitely liked was that the switching enclosures and element centre assemblies came already mounted on the boom. This saved a lot of time and probably eliminates numerous opportunities for assembly errors. All I had to do was bolt the boom together, telescope and mount the tapered element sections, and attach the control and coaxial cables. This antenna would be simple to deploy in the field.

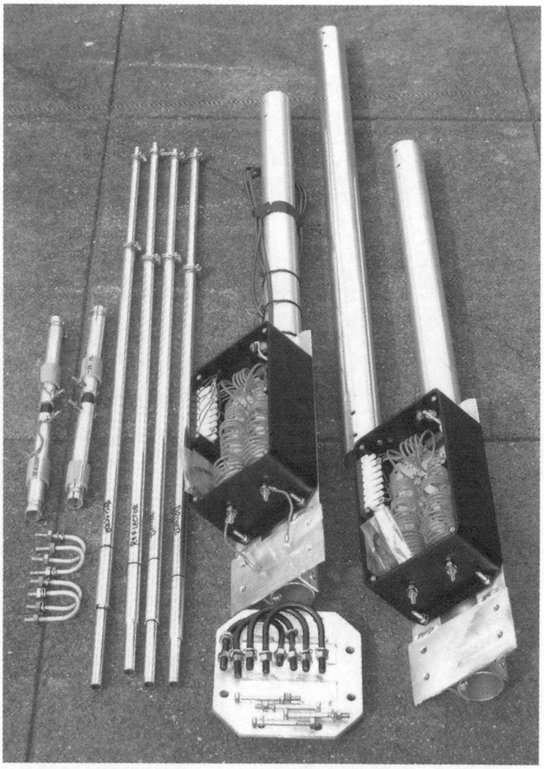

The pieces of the antenna as delivered. All the pieces are 3ft (91cm) or shorter, making the Q-52 easy to transport for portable operation.

Once I started the assembly and tuning process, I found several nits that are typical of new products. For example, the list of needed tools is missing a couple of items and some additional text or a drawing is needed in a few places. It would be a good idea to specify the use of anti-oxidant compound when assembling all of the metal-to-metal junctions, too.

The electrical assembly is very simple – stick the control cable through the hole and attach the wires in a certain order. The only complaint I have with this part of the process is that the terminal strip orientation makes it difficult to tighten the screws but it is not likely this will need to be done on a regular basis. I also had to tin the stranded wires of my control cable to clamp properly with the solid wire of the provided cable. This minor difficulty aside, the checkout routine went smoothly.

The Q-52 is light so it was not hard to carry the antenna up on to the roof for testing and then hold it in place while wrestling the U-bolts into place around the mast. The antenna was easily balanced and mechanically aligned at the mounting point, as well. For a portable installation this would be a big plus. Finding the balance point for the boom-to-mast plate is a step that should be included in the ground level assembly section of the manual.

The antenna was installed not far above a wood and composition shingle roof. I found this affected the tuning of the antenna on 20m quite a bit. The initial frequencies of minimum SWR were about 200kHz lower than expected although higher frequency bands were less affected. After some conversation with Tom, N6BT, a field readjustment procedure was created and has been added to the manual along with an explanation of the circumstances that affect element tuning. In general, any antenna with a stand-off distance of less than 1/4 wavelength from other antennas or surfaces will be detuned compared to the same antenna mounted in the clear. This is particularly true for Yagis and other parasitic arrays that are highly dependent on coupling.

Back on the roof, and starting with the as delivered element lengths and coil configuration, I was able to raise the 20m resonant frequencies easily by adjusting element lengths and the loading coil values. Coil inductance is adjusted by squeezing the turns together (more inductance) or spreading them apart (less inductance).

As you might expect, changing the 20m settings also required some tweaking of coils on the higher bands. All told, it took a bit more than a half-hour of adjustment and checking to get the antenna tuned to where I wanted it. The final settings are given in **Table 6.6**. The two 20m tuning positions overlap enough to keep the SWR below 2:1 from the bottom of the band to about 14.300MHz. My rig's internal tuner easily handled the mismatch above 14.300MHz.

Since the Q-52 is likely to be installed at low heights, near roofs and buildings, and in the vicinity of other antennas, owners should expect to have to adjust the antenna's tuning on the lower bands, at least. Rotating the antenna so that the elements were parallel to my 105 foot 32m) long dipole, which is about 15 feet (4.5m) from the beam, did show some interaction, adding 0.2 to the SWR on 20m. No other bands seemed to be greatly affected and I am using both antennas without any problems.

I also have nearby a 14 foot (4.3m) vertical antenna / auto-tuner combination with the base height a few feet below the beam. I rewired the shack antenna switching so that I could quickly switch among the dipole, beam and vertical for comparisons. HF propagation is highly variable and very dependent on vertical take-off angle so it is challenging to give a true comparison of the beam to the dipole and vertical using live signals. I often found signal levels changing by 20dB or more in a matter of seconds among the three antennas as conditions along the signal path changed.

On 10 and 12m, where weak signals accentuate an antenna's strength and weaknesses, I found the Q-52 and dipole to be competitive in the dipole's most favoured directions and the Q-52 regularly an S unit or two better away from the dipole's main lobes – the manual provides comparison patterns between a 102-foot G5RV an-

Band setting	Frequency (MHz)	SWR	2:1 SWR Bandwidth
20m Low	14.080	1.3:1	140kHz
20m High	14.220	1.3:1	170kHz
17m	18.090	1.3:1	250kHz
15m	21.120	1.5:1	900kHz
12m	24.95	1.1:1	1.5MHz
10m	28.7	1.2:1	1.9MHz

Table 6.6: Q-52 SWR data, with the antenna fed through 50ft (15.2m) RG-8X coaxial cable.

tenna and the beam. The vertical generally came in third although at certain times and in certain directions it would outshine either of the higher antennas. I kept the antenna switch busy!

I found signals on the Q-52 to be up to an S unit stronger on the lower bands than either the dipole or vertical. The Q-52 performed better in directions where the dipole pattern had a null along its axis and on bands where the vertical's gain approached 0dBd. It wasn't that the Q-52 had all that much more gain than the dipole or vertical but being able to aim the gain carried the day several times beaming over the North Pole and straight east or west to Africa and the Pacific on these bands. Sometimes the height of the individual antennas was just right for a particular signal at a particular time of day, too.

The Q-52's ability to reject noise in directions the symmetrical dipole and vertical could not was also welcome, improving signal-to-noise ratio significantly on occasions. Even at the low height of 30 feet (9.1m – about 1/2 wavelength on 20m) the advantages of the Yagi pattern were apparent in rejecting interference from other signals and from local noise sources, of which there are many in this urban location.

The Q-52 is a very attractive option for portable operations, where the Q-52 is very competitive with other small Yagis. The modest amount of directivity improves the receive signal-to-noise ratio for easier copy of marginal signals and the extra gain over a dipole is always welcome.

As of January 2020 Tom Schiller, N6BT, trades under the company name Next Generation Antennas. The Q-52 can be ordered from the www.nextgenerationantennas.com website. It costs $689.

YAESU YA-30 BROADBAND ANTENNA

RSGB Staff

RadCom, December 2012

For portable operations on HF, a half-wave dipole is one of the quickest and easiest antennas to put up. They are simple to make oneself and efficient radiators. The problem, of course, is that by definition they are single-band affairs. What if you want to operate on all bands from 1.8 to 28MHz and don't want to have to put up multiple antennas? The Yaesu YA-30 could be the answer...

Designed by Yaesu to satisfy the commercial and amateur demands for a wideband HF antenna, the YA-30 is a very substantial but compact antenna. Although advertised as a folded dipole it is, in fact, a terminated folded dipole, very similar to the T2FD (Tilted Terminated Folded Dipole) design that is used extensively by commercial HF radio stations.

The T2FD design first appeared back in the 1940s and, after extensive testing by the US Navy, was adopted as their primary wideband antenna. One of the main attractions of the T2FD design is its wideband performance

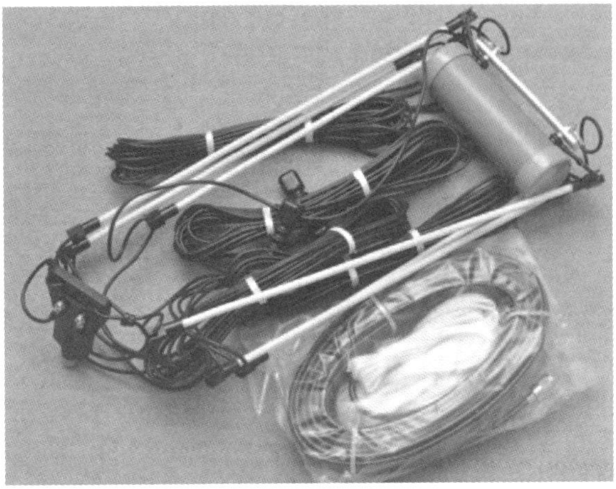

that enables stations to operate across the entire HF spectrum with just a single antenna.

Although not as efficient as an antenna designed for a specific frequency, the antenna is still one of the best compromises available for wideband operation. As such it is an excellent choice for multiband amateur radio stations where only one antenna can be installed.

This T2FD design is often described as a 'squashed rhombic' antenna, as the rhombic antenna is its closest relative. The antenna comprises a folded element similar to a simple folded dipole but the centre of the folded element is cut and a terminating resistor is inserted. The value and construction of this resistor is critical as it has to be able to dissipate power but cannot be wirewound as the inductance would disturb the performance of the antenna. The value of the resistor is determined by the desired feeder impedance and in most cases is just over 300Ω to match a 300Ω feeder.

In the case of the YA-30, the antenna is supplied with a balun to match the feed point impedance to the supplied 50Ω coaxial feeder.

Construction of the YA-30 is very substantial. It is clearly designed for commercial as well as amateur use and, as such, it should be good for many years of service. Like all antennas, the YA-30 ideally needs to be mounted as high and as clear as possible for best performance but you will need solid supporting masts as the antenna is quite heavy. However, there is some flexibility in the installation as it can be erected in an inverted-V format with just a single main supporting mast. The overall length of the YA-30 is 25m (83ft) and it is supplied with everything necessary for erection, including guy ropes. The YA-30 is rated at 150 watts PEP from 1.6 to 30MHz and has a feed impedance of 50Ω. The antenna even includes 30m of 50Ω coax terminated in a PL-259 plug. The VSWR is quoted as better than 2:1 from 1.6 to 18MHz and better than 3:1 from 18MHz to 30MHz.

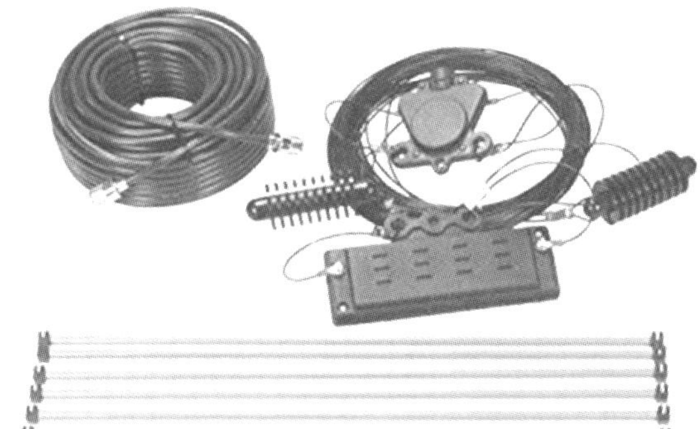

The Yaesu YA-30 is available in the UK from Martin Lynch & Sons (www.hamradio.co.uk) and costs £199.95 inc 20% VAT, excluding shipping.

THE PRO ANTENNAS I-PRO TRAVELLER

Reviewed by Steve Nichols, G0KYA
RadCom, June 2010

Well-known amateur radio writer Steve Nichols, G0KYA, has carried out many reviews of commercial antennas for RadCom over the years, several of them suitable for portable operations. The I-PRO Traveller is a vertical dipole antenna made by Carl Kidd, G4GTW, of Pro Antennas that was designed specifically for portable use. Steve and Chris Danby, G0DWV, braved the north Norfolk coast in the winter of 2010 to put it through its paces...

The I-PRO Traveller is a centre-fed half-wave vertical dipole with capacity hat end loading. Capacity hat loading was chosen to keep the inductive loading to a minimum and so minimise losses. The added benefit is that you have an antenna that can cover 40m – 10m without an ATU, but which stands only 3m tall. The other bonus is the antenna doesn't need any form of ground plane or radials to work.

The Traveller power specification is 1200W PEP on 20m – 10m and 1000W PEP on 40m. The antenna comes packed in a sturdy cardboard box. Once everything is pulled out you are left with a selection of components, including the star-shaped quad-leg base, the two capacity 'hats' and the multi-band centre matching section. The standard band coverage is 20 – 10m, with an optional centre section to provide 40m and 30m coverage.

All of this fits neatly into a 1m long custom built black nylon holdall, which can easily be carried around or put into a car boot. The Traveller also lends itself to explore portable operation overseas as it is easily transported in its compact carrying holdall.

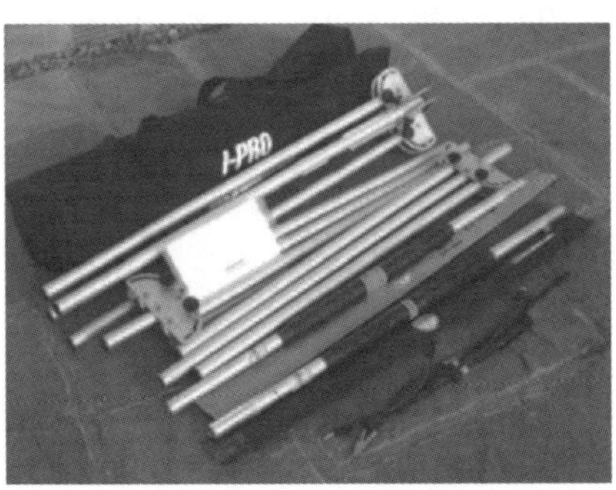

The antenna packs away neatly into its own carrying case.

The antenna comes with a spirit level to ensure that you get it perfectly upright.

The dipole pieces are made of high-quality anodised thick-walled aluminium stock, with right angle bracing and thumbwheels to enable the four legs to be adjusted on uneven ground and the lower capacity hat legs to be moved upwards to tune the antenna – this is especially important on 12m and 10m.

All nuts bolts and screws are stainless steel with plating and anodising protecting bracketing and tubing (which means it could also be left outside in a permanent installation).

The first stage of construction is to set up the quad-legged base. To help with this Carl supplies a small spirit level that plugs into the base and helps you get the whole thing truly upright. The quad-legged base allows the Traveller to be used with the ground sloping by as much as 30°.

Once you have done that you can slot in the lower capacity hat, multi-band centre matching section and top capacity hat all in one go. The capacity hat sections are a slot fit into the centre pieces, with small stainless steel press button clips to lock them into position. We found that the pieces were a very tight fit indeed and disassembling the antenna in cold conditions was quite difficult. A smear of Vaseline or something similar would help no end.

You then fit a small 30cm long 8mm thick fibreglass support piece which supports the coax feed at right angles to the antenna.

Decide which band you wish to use and plug the fly lead in to the gold-plated sockets accordingly – this selects different loading coils on the lower bands. Pro Antenna supply a small laminated instruction sheet with the tapping points shown, which makes life simpler. These tapping points give the rough points for each band. You then tune the antenna to resonance by moving the bottom capacity hat sections up or down, so making them into a V-shape.

The whole assembly weighs around 7kg and can easily be moved around by one person.

Using a Timewave analyser Chris, G0DWV, and I were very quickly able to resonate the antenna on each of the bands from 20m – 10m. The bandwidth of the antenna was very impressive – we were able to get SWR lows of around 1.1:1 in the middle of each band and concur with the I-PRO Traveller specifications that you should be able to get the following 1.5:1 bandwidths:

20m – 300kHz
17m – 500kHz
15m – 900kHz
12m – 1800kHz
10m – 2200kHz

An antenna analyser is not required to set up the Traveller – the built-in SWR metering in most radios will be fine. We were able to use the antenna without an ATU, only having to move the arms up or down to centre the lowest SWR point on the portion of the band we wished to operate.

If you wish to operate 40m you remove the centre multi-band centre matching section and replace it with a dedicated 40m section, complete with loading coils. We were able to achieve a 1:1 match on 40m with about a 70kHz bandwidth between the 1.5:1 SWR points.

The whole structure is quite sturdy and survived gale force winds during the test period without guying, but make sure you fully tighten the thumbwheels that hold the base legs in place.

To give the antenna a fair test we pitted it against a wide range of base station antennas and took it to a very cold and windswept Norfolk beach in January. We first tested it against Chris's, G0DWV, Cushcraft A4S triband beam, which had been lowered to around 25ft due to high winds. We also compared it against Chris's 132ft doublet at the same height. We then tested it against my dipoles and 132ft OCF dipole (Windom) at 25ft and a 65ft inverted-L with 9:1 unun and ground system.

Setting up on a cold and windy Norfolk beach took just a few minutes. This is Chris, G0DWV, checking the SWR.

At Chris's we found that the antenna performed pretty much the same as the 132ft doublet on 20m and 17m. Sometimes it was slightly better and sometimes slightly worse. What was gratifying was that the noise level was lower on the Traveller by about 2 – 3 S-points. The similar reports on the two antennas were confirmed by a contact with a station in Italy who couldn't tell the difference between the two antennas. We also listened on CW to a station in 8P9 (Barbados) and there was no difference. Chris's beam was consistently 2 – 3 S-points better, which was to be expected.

With the 40m centre section fitted Chris was able to take part in our Norfolk Amateur Radio Club's 40m net with a station in Koblenz, one of Norwich's twin cities, receiving 59 reports each way with 100W. In these tests the Traveller was down slightly on the 132ft doublet on 40m, but that is hardly surprising given that the antenna is only 3m tall – a fraction of the full 40m wavelength.

At my station (G0KYA) the Traveller was quite competitive on a closing 20m band. Signals were roughly equal to my dipoles and 132ft off-centre fed dipole, sometimes slightly weaker as propagation varied. The antenna was more than capable of working across the Atlantic to the USA, Canada and the Caribbean. It was around 2 S-points down on my 65ft inverted-L with 9:1 unun with the top at 9m, which is a good DX performer on HF.

On 17m the I-PRO was either equal to, or outperformed, all my other antennas by about 1 S-point.

DX on the water's edge – the whole antenna is very light and can be moved around easily. (It can also be used during the summer months!)

We then took it to Walcott on the north Norfolk coast and set it up right next to the sea's edge. Carl is a keen advocate of water's-edge DXing and we soon saw why. The sea acts as a giant ground plane which, coupled with the zero degree take-off stretching out to the horizon, guarantees Yagi-like performance.

We were soon listening to VK (Australia) stations aplenty on 20m, and had a genuine 57 – 59 short-path QSO with Mike, VK3XL, near Melbourne, straight out across the North Sea. Mike was using a two-element tribander and 200W, but switched to a quarter-wave vertical and we were still able to continue the conversation. That's amazing performance and I had to video the QSO as I couldn't believe it. Mike also said that we were the only station he could hear out of the UK.

We also logged VK3BCY and VK3VBC and could easily hear many of the IBP beacons on 14.100MHz, including LU4AA in Brazil. Unfortunately, 17m (18MHz) was dead.

We then swapped over to 40m and had solid 59+ QSOs with a lot of European stations. All this with an antenna just 3m high.

Both Chris and I are now sold on the enhancement to be had from sea-edge locations and will be returning to the beach – but perhaps when it is warmer!

What did strike me during the tests was how easy the Traveller is to erect and, as it is freestanding, how effective it is without needing high supports or ground radials. This really can be put up on the beach for an afternoon's HF DXing without too much trouble.

I do urge you to try HF operation from the sea's edge – I would not have believed what you can work if I hadn't seen it with my own eyes.

In January 2020 the cost of the multi-band I-PRO Traveller had recently been reduced to £280.80, with the 40 and 30m option priced at £127.95. For more information see www.proantennas.co.uk

PRO ANTENNAS DMV-II 80M ANTENNA

Reviewed by Steve Nichols, G0KYA
RadCom, August 2019

More than nine years on from G0KYA's review of the I-PRO Traveller, Steve took a look at another antenna made by Pro Antennas, the DMV-II, which was designed to make 80m portable operating easy – even if you do not have a lot of space to play with. As a bonus, with optional loading coils it can also work on 60m or 40m...

The DMV-II antenna from Pro Antennas is the latest in a line of portable antennas designed by Carl Kidd, G4GTW, who also produced the Dual Beam Pro, I-Pro Home and I-Pro Traveller. These antennas quickly gained a fine reputation and sales grew around the world. Unfortunately, in 2016 Carl suffered from ill health and the business was taken over by Tony Parkin, M0IVK, in February 2018. Carl has since recovered and continues to provide Tony with technical support.

The DMV-II is a centre-fed 'V' antenna with two loading coils to make it resonant in the 80m band. The intention is to make it easy to set up. I call it a 'V', but it is really an 'M', as the two 5m telescopic fibreglass poles support the wire element, but then it dangles down at the ends.

The antenna arrived from Pro Antennas very well packed and its components look like top quality. The first task is to assemble the mast-boom clamp assembly to the white fibreglass mast support, which consists of a selection of clamps on two studs with wing nuts.

The antenna is 'M' shaped: the two ends can be seen dangling down from the top of the fibreglass rods.

Make sure you wear gloves to protect yourself from the fibreglass mast.

The instructions warn you to use protective gloves, as shards of fibreglass can otherwise embed themselves in your hand and cause irritation. It is a good idea not to ignore this, as I know from previous experience that otherwise you could have pricking sensations in your hands and fingers for days!

The clamp assembly proved fiddly, rather than difficult, to assemble, but the instructions were quite clear. Once you have built this you slip the feedpoint box on to the short mast support. The instructions say that you only have to do this once – it doesn't have to be taken apart as it can be transported to another location in one piece.

You then extend the fibreglass fishing poles and give each section a gentle twist to hold them in place. You can then slip them on to the aluminium support pieces, having removed the screw-on end caps.

You fix the 80m loading coils to the fibreglass fishing rods with the supplied elastic toggles. These are easily fitted and hold the coils firmly in place. Having done that you can add the colour-coded antenna wires to the coils, sliding the supplied small rubber tubing over the end of the fishing poles to keep the wire taut. You also get plastic clips that hold the wire to the pole, these plus a couple of twists of the element around the pole are enough to keep it from sagging.

You can then add the secondary elements that fit between the loading coils and hang down vertically from the end of the fibreglass pole.

One thing that confused me was that the antenna elements were of unequal lengths. A quick email exchange with Tony confirmed that this was correct – he said the antenna uses a slightly offset feedpoint to help with matching.

For a permanent installation the instructions say that PVC tape can be used to fix the wire to the poles, although you would also need to provide some sort of waterproofing to the plugs as well. I recommend using PVC tape as otherwise the plugs can pull out of the loading coil as the whips move around as you push it into the air.

It is up to you as to when you want to mount it to the (not supplied) mast. I found it was easier to build the complete antenna first before bolting it on, but your mileage may vary.

The mast clamps can accommodate anything from 32 – 50mm diameter, but I would recommend a lighter mast if possible as it can be a little unwieldy once you try to lift it into the air, despite only weighing 3kg.

I would also recommend a ground screw (not supplied) and guys. Another pair of helping hands would also help, as trying to balance the antenna assembly while trying to tighten the guys can get a bit hairy – especially in windy weather! You can do it on your own, but it isn't easy.

To lower the SWR the antenna comes with pairs of 50, 100, 200 and 400mm extension leads, which can be 'cascaded' to give any length of up to 750mm in 50mm increments. The instructions say that as a rule of thumb adding 50mm to each side will decrease the resonant frequency by around 20kHz, so theoretically this could give you the ability to lower the resonant point by up to 300kHz, effectively covering the whole band. It is best to measure the SWR with the antenna at the height you will operate it, as the resonant frequency will shift upwards 10 – 20kHz as you raise it.

I set the antenna on a guyed portable mast at a height of about three metres. The antenna's instructions said it is designed to be resonant at the

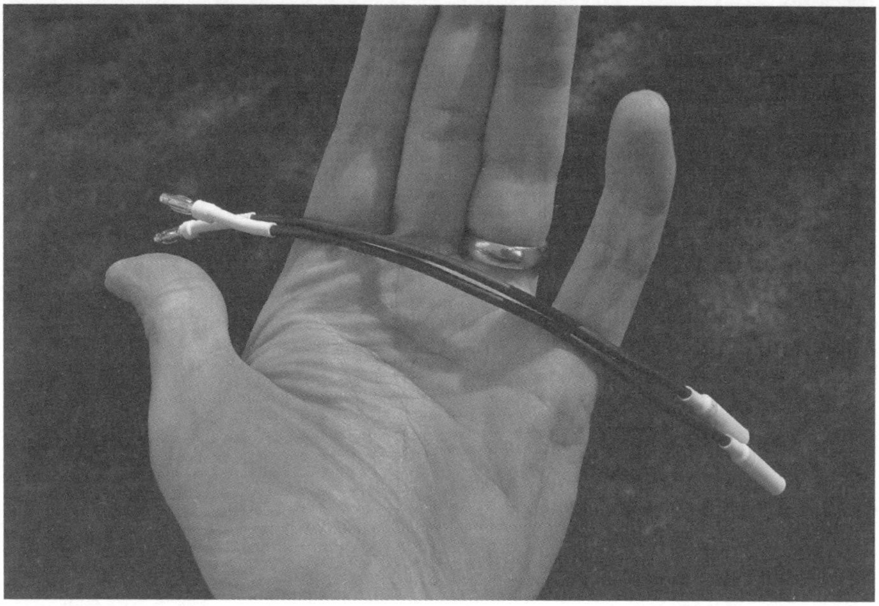

The antenna comes with assorted 'extenders' to move the resonant frequency downwards.

top end of the 80m band and I found that it had a minimum SWR of 1.2:1 on 3.800MHz, which tallies with the instructions.

I then added the 200mm extensions to the ends and the resonant point moved down to 3.690MHz, with a 3:1 bandwidth of 90kHz (3.630 – 3.718MHz).

Adding a further 400mm took the resonant point right down to 3.515MHz – just right for CW (3:1 points below 3.500MHz to 3.554MHz).

Removing the 20cm and just adding the 400mm extensions took the resonant point to 3.600MHz (3:1 points 3.556 – 3.639MHz).

This suggests that an ATU would be a necessity. In fact, it is better if you set it up for the section of the band you require – CW, data or SSB, and just use the ATU to tune out the slight mismatch, perhaps up to an SWR of 3:1.

I started listening on the band at 1500UTC in May, which is not the ideal time to go DXing on 80m! Nevertheless, there were stations and around the UK signals were either equal to or no more than one S-point down on my W5GI dipole.

Noise was either equal to or down a couple of S-points on the W5GI, reflecting the fact that the DMV-II was situated further away from the house.

Extensive back to back tests with *WSPR* across Europe showed that the DMV-II was on average about 9dB down on the W5GI, which is amazing considering its small size.

A contact with John, G8MNY, in Croydon on 3.710MHz showed that he couldn't detect any difference in signal strength between the DMV-II and the W5GI. I had a solid QSO with John for about 30 minutes.

Other QSOs showed it was either equal to or slightly down on the W5GI antenna.

The antenna tested came with two optional loading coils (colour coded blue) that are wound to provide resonance on 5MHz. These slotted into place on the poles and use the existing antenna wires.

Putting the mast back up I found that these gave an SWR of 1.5:1 at 5.350MHz (3:1 bandwidth of 160kHz, 5.335 – 5.495MHz). The wire extensions can then be used to lower the resonant point, much the same as when the antenna is used on 80m.

The Volmet signal on 5.505MHz was romping in at S9+20dB, but try as I might, I couldn't find any amateur activity on 60m. Nevertheless, the figures suggest the antenna should work as long as it is tuned properly on the frequency of interest.

The DMV-II 80m antenna is portable and gives you access to 80m with a 'wingspan' of only 8 metres. This is unusual as a half-wave dipole for the band would be around 40m (135ft) long and a compromise 80m antenna, such as a G5RV or W5GI 'mystery antenna', would be at least 31m (102ft).

There is no doubt that you lose some performance with such a small antenna, but that is missing the point. The fact that the antenna can be folded up and put into a car boot in seconds, and then re-erected in a field or on a beach in minutes is a boon.

The DMV-II is ideal for use as a portable antenna on 80m or, with the optional loading coils, on 60m or 40m.

It might suit people with motor caravans who would otherwise miss out on 80m when out and about.

The problem with using a shortened loaded dipole is that you do have limited bandwidth and need to set the resonant point where you need it before operation. I therefore recommend you make a note of the SWR figures and resonant points with each extender. The instructions handily give you a chart to note these down.

Don't rely on an inbuilt automatic ATU to cover the whole band without adjusting the element extensions as it is unlikely to have sufficient range – mine certainly didn't.

The materials used in its construction are of a good quality and the use of heat-shrink tubing on all the supplied gold-plated plugs was a nice touch.

If you are into portable operation the DMV-II would make a good purchase. Or if you want to operate 80m from home, but don't have the space for a 100 – 132ft dipole, then it will get you on the band. But if you do this, do take care to ensure you have waterproofed all the joints and use PVC insulation tape instead of the toggles and elastic to affix the loading coils.

According to the Pro Antennas website, "the telescopic arms fold down to less than 1.2m for transportation which makes it ideal for use on location... Total weight including mast clamp: 3.0kg. Maximum peak envelope power: 500W." In January 2020 the DMV-II cost £214.20 (recently reduced from £238) and it is available from Pro Antennas (www.proantennas.co.uk). The optional 60m and 40m coils are approximately £40 per pair.

THE SUPER ANTENNA MP1B 'SUPERSTICK' PORTABLE ANTENNA

Reviewed by Steve Nichols, G0KYA

RadCom, January 2013

Back in 2013 Steve Nichols reviewed the MP1B ' SuperStick', which remains a popular antenna among portable operators to this day. The reason: it works on pretty much all bands and can be installed more or less anywhere in no time at all!

When summer is here many amateurs' minds turn to outdoor portable operation. In my own group, the Norfolk Amateur Radio Club, we even have an event called 'Radio by the Seaside' when we all decant to the North Norfolk coast for some HF fun. In fact, we often get some surprises with regular contacts outside Europe each year, thanks to the excellent take-off from the cliff tops.

But the perennial problem is always 'what antenna should I use?' when portable. Ideally, you need something lightweight, compact, multiband and efficient – while you may get any two of these attributes, the likelihood of getting all four is usually remote!

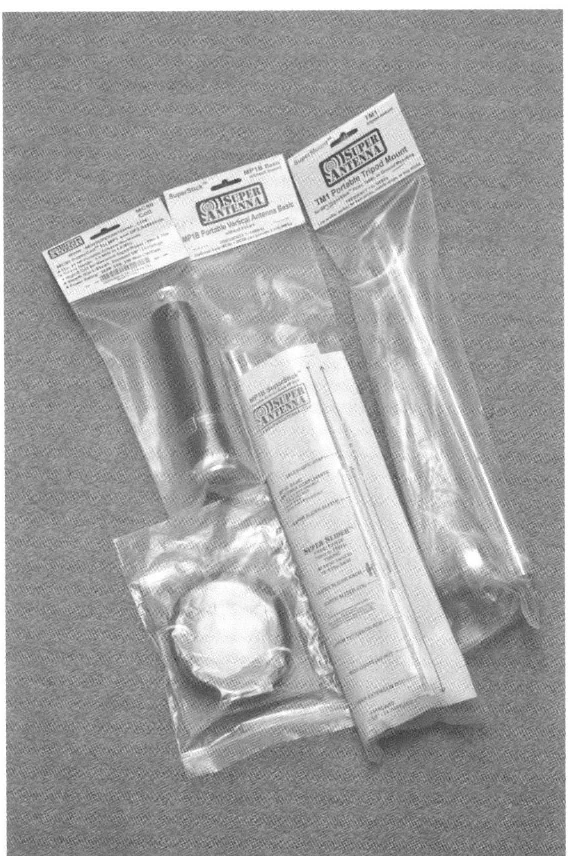

The whole antenna weighs less than 1kg and will easily fit in luggage.

Martin Lynch and Sons think they have the answer with the Super Antenna MP1B SuperStick portable HF antenna, the latest iteration of the original design begun in 2000 by Vern Wright, W6MMA.

The basic MP1B (without the additional 80m loading coil) is a 10 – 40m portable vertical antenna, ideally suited to compact HF rigs (FT-817 / IC-703) and also, because of its 500W SSB or 300W CW and digital modes power handling, it can also be used with more powerful HF transceivers.

When the postman arrived with the package containing the MP1B I was quite surprised. The box was tiny – in fact it was about 3.5 inches square and 15 inches long (9cm square x 38cm long). And it was quite light too – tipping the scales at a little over a kilogram.

On opening the box you are met with an assortment of alloy rods, a telescopic whip for the top of the antenna, radials and a tripod mount. The full contents are as follows:

- One SuperSlider, variable sliding SuperCoil
- One telescopic SuperWhip tip
- Two extension bottom rods with one coupling nut
- One basic counterpoise MR1 system
- TM1 SuperMount low-profile tripod
- A free MC80 coil for 80m operation

The instructions come on a double-sided sheet of A4 paper. To be honest these are a little minimalistic, but I was able to put the whole antenna together as you really can't go far wrong. The instructions show that you can use the antenna anywhere from 3.5 – 440MHz by adding or removing sections.

Also enclosed was a separate sheet that outlines the other accessories you can buy, including the various 'Superwire' radial sets that are pre-tuned for bands from 80m – 70cm; the RM1 SuperMount for fitting the antenna to the side of a rig, such as the Yaesu FT-817 or Icom IC-706; the UM1 SuperMount for clamping to tables, fences, railings and the like; plus the MC60 loading coil for 5MHz use.

The key to the antenna is a variable inductance coil. To use this you slide the top section up or down the coil to short out the turns – a small sprung contact moves up and down the coil as you move it. This is clever as by using this method the instructions say you can tune from 40m (7MHz) to 4m (70MHz).

The instructions give you a guide as to where you need to have the coil set for each band – ultimately you are going to have to use an SWR meter or antenna analyser to get it spot on. Note there are no markings on the coil itself so a little guesswork is needed. Once you have the correct setting you tighten a small nylon wing nut to hold the coil in place.

For 7 – 70MHz use you use both of the aluminium supports rods and the adjustable coil to bring the total length of the antenna to 7.5 feet (2.28m).

If you want to work VHF / UHF (144 / 430MHz) you just use the telescopic top section. For 60m (5MHz) there is an optional (MC60) loading coil. And for 80m use you use everything – the two alloy extension rods, the 80m loading coil, the adjustable coil and the telescopic top.

Close-up of the MP1B's variable inductance coil.

A small rig like the Yaesu FT-817 with the Super-Stick means you can operate from almost anywhere.

The net result is a very versatile package that means that you only need to carry one antenna to work nearly all the most common HF, VHF and UHF bands.

The finish is polished alloy for the tripod and main tuning coil and the antenna look very well made. I was warned that I had a pre-production model and might have problems screwing in the tripod legs and this was the case, although they did fit with some care. I understand the problem has been fixed with the production models.

But what you really want to know is how well does it work? Setting up the antenna in my back garden with the four supplied radials took about five minutes. The three bundled radials are different lengths – 2.5m, 3.5m and 5m. These are pre-tuned quarter waves for 10m, 15m and 20m (allowing for the velocity factor of PVC-coated wire) although lying them on the ground will detune them. These were easy to deploy, but if you want to experiment you could add more and longer radials – probably best on the lower bands. As I said earlier Super Antenna can supply different packs of pre-tuned radials for a wide range of bands, or you could make your own. Having laid all this out I then spent around 30 minutes going from band to band to see what the lowest SWR was on each band and what the bandwidth was.

Starting with 6m I was unable to get a low SWR reading, until I realised that you have to shorten the telescopic section. Once I did this I was able to get a lowest reading of 1.4:1 on 50.100MHz. The bandwidth was excellent but unfortunately there were no stations on the band at that time.

Moving on to 10m, I extended the telescopic section and took a reading with my MFJ antenna analyser. After moving the coil up about four clicks I was able to get an SWR reading of 1.2:1 on 28.480MHz. It was very broad-banded with a reading of 1.3:1 at 28.000MHz and 1.5:1 on 29.7MHz.

The SWR on 12m (24MHz) was equally low – 1.3:1 across the entire band. Pulling the coil up a little more and 15m (21MHz) was also very acceptable at no more than 1.6:1 across the entire band.

On 17m (18MHz) the lowest SWR was 1.5:1 and on 20m I found that you really need to tune it for either the CW end of the band or the SSB section.

It was only one or two clicks to move up or down the band, but this meant that you were able to operate with an SWR of 1.6:1 wherever you were on the band.

It became apparent that each successive band change needed more movement of the coil, which is to be expected. On 30m (10MHz) the antenna could cover the whole band at around 1.4 – 1.6:1. On 40m while the antenna could be set up to give a 1.1:1 match on 7.1MHz it had risen to 2.6:1 at the band edges, so again it is better to choose the area of the band you wish to work and set the antenna up for that portion.

If I were to own the antenna I would make up a small tuning guide, photocopying the guide on the instructions and encapsulating it.

Once I had finished the SWR tests it was time to see how it performed. After all, a low SWR is no guarantee that an antenna will work on the air.

The 10m band was humming with Sporadic E (Es) activity and the EA4Q and F5ZUU beacons were quickly logged – both only running 5W according to their CW idents. I made contacts with Marco, IZ5RVG, who was a solid 559 on CW and Rudy, HB9CCL, who was a solid S8 on SSB. Compared with a full size 10m half-wave dipole at about 30ft the MP1B was about one or two S-points down, which is to be expected given that it is only about seven feet long. But some signals were exactly the same strength – I had a 10 minute SSB QSO with Sebastian, DL3GAZ, during which he couldn't see any difference in signal strength when I switched to the longer antenna.

I spent some time on 17m and was easily able to make CW contacts on the band with 100W. For example, I received the obligatory '599' from Igor, EW8A, in Belarus on the first call using the MP1B. In fact, Hartmut, DL1VDL, was the same strength on the MP1B as he was on a full-size dipole for 17m.

It was still quite impressive on 20m where signals were down a little compared with a full-size dipole, but louder stations were still easily workable. Again, there were some stations that were the same strength. As you slide down in frequency you may find the modes that are more suited to weak signal work, such as CW and PSK31 [or now, in 2020, FT8 and FT4 (which had not been invented when this review was written!) – *Ed*], give you the best results. On 7MHz (40m) the antenna was down about 20dB on a dipole, which is not surprising.

On 80m it was even worse – you won't get far with QRP on these bands and you may struggle with 100W as well.

I took the MP1B to the Norfolk Coast where I made many HF contacts using both SSB and CW, mainly on 10m using my Yaesu FT-817 and 5W. The ease with which the antenna can be deployed means you can be up and running in minutes.

The antenna gets good reviews on eHam.net and one happy buyer even sent the following to Martin Lynch after purchasing one at the RSGB Convention: "In less than ideal conditions with the SuperStick strung up from my curtain pole on the inside of the flat window, I was amazed to find

The SuperStick can be assembled in minutes. It is seen here on the North Norfolk coast.

that in just one weekend I managed to work the US, Canada, Jamaica, Colombia and Brazil. I'm very much looking forward to trying it out outdoors but I can honestly say that this works well as an indoor stealth antenna. So far I've managed to manually tune to almost 1:1 SWR on 20m and 40m and 1.5 – 2:1 on 15m, though I still had no trouble getting a hundred contracts on 15m during CQWW."

The antenna is obviously at its most efficient on the higher HF bands and this really showed on 10m. If you are serious about portable operation I would also recommend experimenting with more, longer radials – especially on the lower bands.

All in all then, the SuperStick MP1B is a lightweight, simple antenna system that breaks down into a tiny package, yet still offers reasonable performance on the higher HF bands. It is a real compromise on 80m, but then what do you expect given that it is only seven feet tall?

Our thanks to Martin Lynch and Sons for the loan of the antenna.

Although the basic design is similar, the SuperStick MP1B has evolved since this review. The latest versions are available in the UK from Martin Lynch & Sons (www.HamRadio.co.uk), starting at £249.95. See the ML&S website for the options available.

SPIDERBEAM AERIAL-51 MODEL 404-UL

Reviewed by Steve Nichols, G0KYA
RadCom, May 2015

Steve Nichols reviews a lightweight 20m-long off-centre fed dipole made by the German Spiderbeam company, famous for its portable wire Yagis and fibreglass antenna support poles.

Portable and holiday-type HF operation seems to be gaining in popularity. This may be helped by the appearance of lightweight radios like the Yaesu FT-817 and Elecraft KX3. And if QRP isn't your scene, transceivers like the Icom IC-706 or Kenwood TS-480 can give you 100W all-band performance in a transportable package. The only problem is that you still need an antenna to use when you get to your dream island location.

Rick, DJ0IP, at Germany's Spiderbeam, with its Aerial-51 brand, has come up with a possible solution with its lightweight 20m-long Off Centre Fed Dipole or OFCD. (I think Aerial-51 is a pun on the infamous top secret air base in the USA – think about it!)

Weighing 400g including balun and 12m of RG-174-50 low-loss coax, the Aerial-51 Model 404-UL is an ultra-lightweight, asymmetrical dipole operating on seven or eight shortwave bands, namely 40m, 20m, 17m, 15m, 12m, 10m and 6m with an SWR of less than 4:1. It will also operate on 30m at less than 6:1. Spiderbeam says the supplied coax offers lower losses than the more common RG-174U, with a worst-case scenario on 28MHz of –1.61dB and –2.17dB on 50MHz.

You should aim to use an ATU on at least the 30m, 17m and 12m bands, and one may also be useful on the other bands in order to bring the match right down to 1:1.

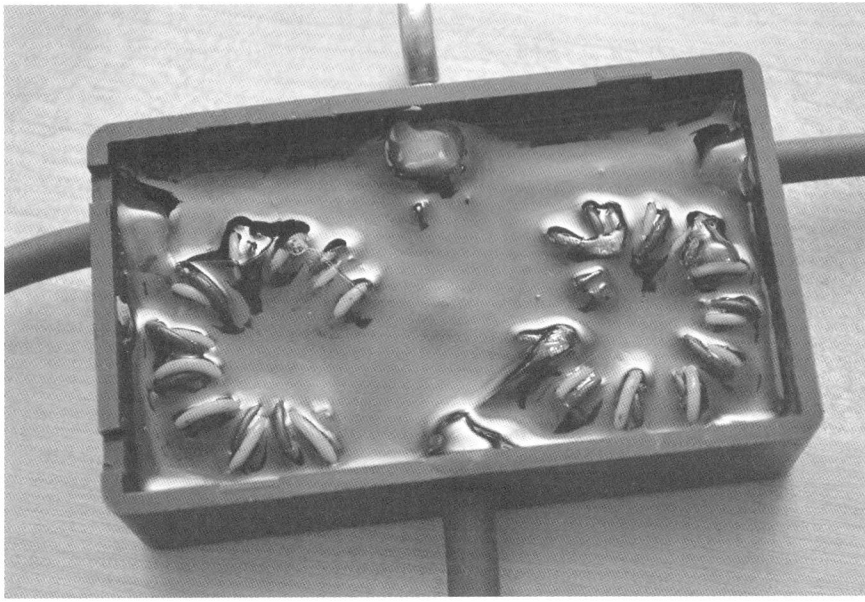

The balun uses two matched toroidal transformers potted in epoxy.

The OCFD is rated for a maximum 200W power, apart from 30m where power should be restricted to 50W to avoid overheating the balun.

The antenna is designed to be strung in an inverted-V style using a single support, such as a telescoping fibreglass fishing pole. Spiderbeam sells these in 12, 18 and 26m variants, although the latter two cannot really be classed as being suitable for suitcase-type operations.

All the hardware used is stainless steel and the antenna wire is made from CQ-534 insulated 1mm OD multi-strand copper-clad steel wire, which has very low wind load and weighs only 31g, yet has a breaking strength of 10kg.

The lightweight 4:1 ‘Guanella’ current balun matching unit uses two matched toroidal transformers potted in epoxy for rigidity and weather protection. The balun also provides choking resistance to suppress common mode currents. The balun has a stainless steel eye to help with mounting.

The design is a classic off centre fed dipole with two legs – I measured one leg at 8.4m long and the other leg 12.2m, making a total of 20.6m. Both dipole ends finish in small plastic toggles to which supplied heavy-duty monofilament fishing line is attached. This can easily be removed if you want to attach your own supporting cord.

An OCFD is usually fed at the one-third / two-thirds point where the impedance is close to 200 or 300Ω on a number of bands, depending on the antenna’s height above ground. When fed with a 4:1 or 6:1 impedance transformation balun the match is then brought closer to 50Ω. This way, an antenna cut as a half-wave dipole for 40m can usually be persuaded to work on at least 20m and 10m as well.

This dipole is fed at the 59% / 41% point, rather than the more conventional one third / two thirds. The reason for this I think is simple. Modelling the antenna in *MMANA-GAL* shows that when erected as an inverted-V the 59% / 41% feedpoint is better for giving a reasonably low impedance match on all of the bands it is designed for. If it had been fed at the 66.6% / 33.3% point I doubt you would get such a good match on 17m and 15m.

My MMANA-GAL model shows the antenna should offer a reasonable match on 40, 20, 15 and 10m, with a poorer match on 12 and 17m, after adjusting the lengths of the modelled wires to allow for the estimated velocity factor of the PVC-coated wire.

The antenna model shows the antenna tends to be a bit of a cloud-warmer on 40m, with largely Near Vertical Incidence Skywave (NVIS) radiation when mounted as an inverted-V with the apex at about 8m. This is ideal for close-in contacts rather than DX. On the higher bands the pattern breaks up into a multi-lobe shape offering both high and low-angle radiation. If you want to work DX the higher you can mount the antenna and its ends the better. This is where Spiderbeam's 12m pole comes into its own.

7.000MHz	2.1:1
7.100MHz	2.2:1
7.200MHz	2.4:1
10.100MHz	6.6:1
14.000MHz	2.4:1
14.175MHz	2.2:1
14.350MHz	2.0:1
18.068MHz	3.6:1
18.168MHz	3.6:1
21.000MHz	2.4:1
21.225MHz	2.3:1
21.450MHz	2.4:1
24.900MHz	4.1:1
28.000MHz	1.2:1
29.000MHz	2.0:1
29.700MHz	2.6:1
50.000MHz	2.3:1
52.000MHz	2.6:1

Table 6.7: SWR readings at the end of supplied coax.

Once I put the the antenna up at 8 metres I did some SWR tests at the end of the supplied 12m of coax and the results are in **Table 6.7**. These were in line with the manufacturer's claims, although the 10MHz figure was slightly higher. No doubt mounting the antenna a little higher and / or supporting the ends a little higher would bring this within the specification.

The wire used is so thin that it is almost invisible. Although designed for portable operation, if mounting the antenna permanently I would consider using springs or bungees on the ends to remove some of the strain due to wind.

The antenna itself performed very well and was a good competitor to the half-wave dipoles it was tested against. Signals were usually absolutely equal to those on a 40m centre-fed dipole, which is to be expected.

On the higher bands it was a little more of a lottery depending upon whether the station you were working was on one of the antenna's radiation lobes or not. Signals were either equal to the dipole or up or down a little, but this is due to be expected due to the more complex radiation pattern of the OCFD as it is longer than a half wave at these frequencies.

When mounted away from the house, and with a choke balun at the bottom of the feeder, it was a quiet antenna too.

Overall then, I thought the antenna was well built, lightweight with a lot of attention being paid to the quality of the materials and the balun. Feeding the antenna at a slightly different point has resulted in a better match across

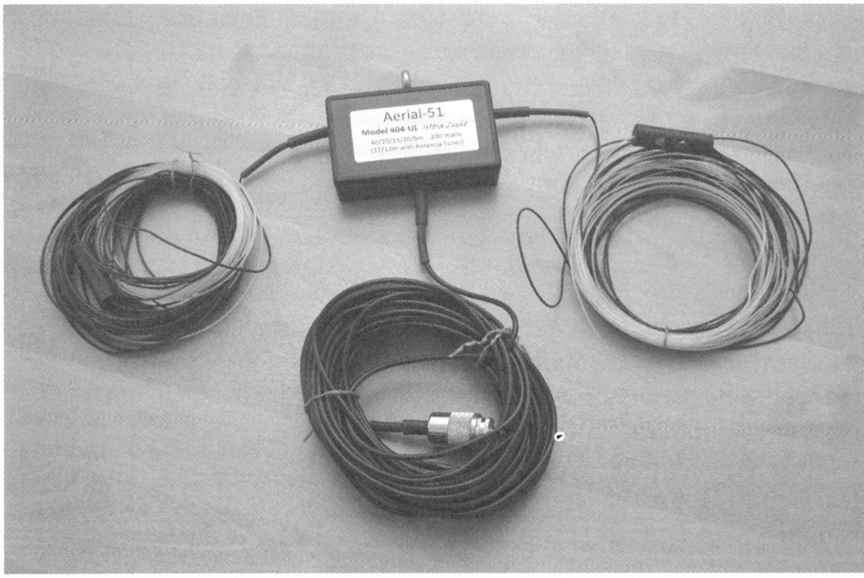

The 404-UL OCFD fulfils its design objective of being a lightweight multi-band HF antenna.

a lot of bands, although I would still recommend using an ATU where possible. An internal ATU would probably have no problems matching the antenna fully on its designed bands, with perhaps only 10MHz being a little more problematic on some makes of radio.

The 404-UL OCFD fulfils its design objective of being a lightweight multi-band HF antenna, suitable for taking on DXpeditions and holidays. I can't help thinking it would make a good low-profile permanent antenna too.

Our thanks to Rick, DJ0IP, at Aerial-51 for the loan of the antenna used in the review. You can find out more at www.aerial-51.com

The Aerial-51 404-UL OCFD antenna is manufactured in Germany by Spiderbeam (www.spiderbeam.com) and it is available from them for 99 euros inc. 19% VAT, or 83.19 euros for export outside the EU. In North America it is sold by Spiderbeam USA (www.spiderbeam.us) for US $129.

THE ALPHA ANTENNA 'ALPHA LOOP' PORTABLE MAGNETIC LOOP ANTENNA

Reviewed by Steve Nichols, G0KYA
RadCom, September 2014

The final portable antenna we have selected that was reviewed by Steve Nichols in RadCom is a magnetic loop. He described it as "A compact loop antenna that has a lot to offer". The version reviewed was rated at 30W PEP, making it ideal for the Intermediate or Foundation licensee wanting to give portable operation a try, though the latest version is rated at 100W.

The Alpha Antenna MicroTune magnetic loop.

Portable HF operation seems to be getting more and more popular. While a decent HF antenna traditionally tends to be quite large, whether it is a vertical or a wire dipole that needs some supports, conversely magnetic loop antennas tend to be quite small. Alpha Antenna from the USA had adopted the approach of producing a lightweight magnetic loop (or more correctly a small transmitting loop, STL) antenna that comes in its own carrying case, tipping the scales at just 6.6lb (3kg). This is, therefore, ideal for holiday-type operations or for carrying in the boot of the car for portable operations. Many Alpha Antenna customers also use it where outside antennas are difficult to install or where they are restricted.

The Alpha Loop is a magnetic loop antenna that is easy to tune, has acceptable bandwidths and offers coverage on the 40, 30, 20, 17 and 15 metre bands. Alpha Antenna also includes a small feed loop that has been optimised for 12 and 10 metres, for use after you make a small alteration to the larger loop.

Above: The tuning box has a capacitor with a 6:1 reduction drive and a useful scale plate around the knob.

Left: The feed loop.

The antenna comes complete with a tripod that can be extended to about 2m.

The antenna appears to your transceiver as a large resonant circuit, consisting of a large single turn inductor tuned with a single-section variable air-dielectric butterfly capacitor. It is rated at 30 watts PEP SSB maximum [the latest version is rated at 100W – *Ed*] and it uses a 6:1 reduction drive to make tuning easier.

A nylon shaft has been placed between the knob on the front of the Alpha Match and the capacitor to prevent RF burns and minimise your body's capacitance interfering when tuning the antenna. The Alpha Match is placed inside a high voltage Underwriters Laboratory-rated outdoor waterproof housing.

The antenna comes in a very smart nylon holdall for its field bag, which is just under three feet (87cm) long. The holdall, which has the manufacturer's name and logo neatly embroidered on the side, includes a zip-fastened side pocket that is handy for carrying instructions. The pocket also contains a shoulder strap that can be affixed to the holdall.

The supplied instructions are on two A4 pages, folded to give an eight-page A5 booklet.

Open up the holdall and you'll find a three-section, powder-coated tripod that can be extended up to more than six feet (2m). You'll also find the matching box, with a dial and knob to adjust the capacitor, the feed line box (complete with SO-239) and a bag with curved aluminium strips that make up the loop.

Assembly is quite straightforward. You first affix the small aluminium loop to the feed line box with four supplied stainless steel bolts and wing nuts. This is the main feed loop, which is made from half inch x 1/8-inch (about 12 x 3mm) bar and with a diameter of about 8.75 inches (225mm). Having done that, you then connect the larger radiating loop to the feed point assembly, again using the supplied bolts and wing nuts. The radiating loop is made up of six individual pieces of half inch x 1/8-inch bar that are bolted together. This gives you coverage from 40 – 15m (7 – 21MHz). The end result is an ellipse about 33 inches (84cm) in diameter.

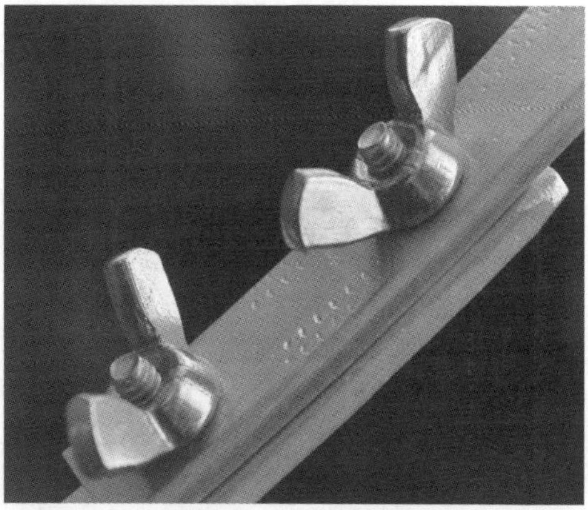

The loop is constructed from curved sections that are connected together using bolts and wing nuts.

If you want to work 12 – 10m (24 – 28MHz), you have to take out two of the longer element pieces that make up the larger outer element. You also have the option to change the inner feed loop for the smaller one supplied, which will optimise SWRs on these bands.

You are given more than enough plated wing nuts and bolts to assemble the antenna.

To make a magnetic loop antenna as efficient as possible you need to reduce its ohmic losses. That is, elements should be made from copper or aluminium and all joints should be brazed or welded to keep the losses to a minimum. Alpha Antenna also offers a magnetic loop antenna with minimal ohmic losses made of two continuous solid elements called the Continuous Alpha Loop.

With this antenna, however, the main loop has up to eight joints (held fast with nuts and bolts) plus there are two further joints in the feed loop. Quite how these joints, and their associated losses, affect the overall performance of the antenna is hard to say. The loop elements overlap by just over an inch and half (40mm), which may help a little.

Once completely assembled, the antenna mounts on the included tripod by screwing it securely in place on a ¼ x 20 thread on the tripod. The whole assembly is reasonably sturdy, but it might be worth guying if using it outdoors in windy weather.

At this point you can attach your coax to the feed point at the top of the loop and tune the capacitor using the knob and its associated pointer. The coaxial feed point is outside of the plane of the antenna, to keep the voltages inside the loop from interacting with those in the coax. I would recommend tuning the antenna for maximum received noise first, which will put you in the ballpark area for resonance.

Note there is a warning label on the tuning box that warns you not to touch the antenna when transmitting as you may receive an RF burn. As someone who has grabbed hold of a magnetic loop antenna when transmitting

40m (7MHz)	25kHz
30m (10MHz)	38kHz
20m (14MHz)	54kHz
17m (18MHz)	82kHz
15m (21MHz)	110kHz
12m (24MHz)	105kHz *
10m (28MHz)	135kHz *

Table 6.8: Bandwidth (between 3:1 SWR points). *Note:* The above were measured at the feed point. Typical operating bandwidths will be slightly larger due to coax losses. Measurements marked * were taken with the smaller feed loop and reduction of the main loop.

100W, I can attest to this: you'll probably only do it once! It also warns you not to operate the antenna if you have a Pacemaker or to stand too close when transmitting.

With the transmitter switched to low power, transmit and peak the tuning capacitor for the lowest SWR. As you may have gathered, this is not a 'fix and forget' antenna and on some bands you will have to tune it to resonance every time you change frequency by more than a few kilohertz.

Specifically, the manufacturer shows the Alpha Loop is able to provide a bandwidth of approximately 20% on 10 metres, 37% on 12 metres, 35% on 15 metres, 85% on 17 metres, 20% on 20 metres, 100% on 30 metres, and 20% of the 40 metre band while maintaining an SWR of between 1.1:1 and 3:1. The measured bandwidths between the SWR 3:1 points are shown in **Table 6.8**.

The instructions suggest that, if you are using an antenna analyser, it might be an idea to have an A/B antenna switch so that you can quickly move from the analyser to the radio when tuning. An antenna analyser, however, is not absolutely necessary as most rigs often a built-in SWR meter. Needless to say, you need to have easy access to the antenna as you will be tuning it a lot. It is also not designed to be left outside all the time – this is strictly a portable antenna and not designed for permanent outdoor installation.

Using an MFJ analyser I found that the antenna would tune from 6.945 to 22.860MHz, achieving a near 1:1 SWR at both the bottom and top ends of its tuning range. Tuning is *very* sharp indeed, which is what you would expect from a magnetic loop, even with the 6:1 reduction drive. To help, the tuning dial does have some markings to help you get to a known rough tuning point quite quickly.

There was some evidence of hand capacitance too, meaning the antenna de-tuned slightly when you took your hand away. This is indicative of an efficient magnetic loop that has a high Q. What also soon became apparent was that tuning was tricky – you need to be quite close to the antenna to tune it, but you also need to be able to key the radio and see the SWR meter. Having an analyser on a switch, or sticking to known frequencies, such as the QRP calling frequencies, would make it easier.

The antenna is directional in the plane of the loop, with which you can peak signals and also null out interference, which is useful when around high RFI environments.

The next stage was to move the antenna outdoors and try it there. I found the easiest way to set it up was to listen and 'peak' the capacitor in the middle of a desired band segment, such as the QRP calling frequency

Another view of the Alpha Antenna portable Alpha Loop.

or near to some loud stations on SSB. This saved having to make minute adjustments each time you QSY. Setting it on 7.030MHz I found that signals on 40m were down about 2 – 3 S-points on a dipole, although the mag loop is a very quiet antenna and signal-to-noise ratios were often very similar.

Setting it on 14.070MHz (PSK31), the antenna was again down about 2 S-points on a dipole, but capable of making contacts. It was a similar story when using CW on 20m. Lots of CQs and few replies, although I think a lot of this was due to the summer HF doldrums and poor band conditions. On SSB, I needed to find someone who was at least S7 to S9 to be able to work them with my FT-817's 5 watts. You will probably have more luck on CW and PSK31 [and, these days, on FT8 / FT4, which had not been invented at the time of this review – *Ed*].

The antenna's performance on 21MHz was a little better, which is to be expected as it is at its most efficient on this band. It was only down about 1 S-point on the reference dipole antenna. I was able to have PSK31 QSOs with the louder stations, such as OM3ZAS.

I then took two of the longer segments out of the loop as instructed to set it up for 10m and 12m. This gives a loop with a diameter of about 21.5 inches. I also changed the feed loop to the smaller one as instructed. The lowest SWR on 12m and 10m was about 1.5:1.

Tuning around 10m during a Sporadic E opening found signals in the S8 to S9+ range. The antenna was about 5 – 10dB down on a dipole, but with signals that strong it didn't really matter. Contacts were quite easy, even when using SSB.

The Alpha Loop antenna is well built, comes in a nice nylon holdall and is easy to assemble with no tools required. It can be assembled in less than 10 minutes and works from 40 – 10m (once you have removed two of the aluminium segments and swapped the feed loop). Tuning is easy, but very sharp and hand capacitance effects are evident.

In terms of its performance, it will work better for you on CW and datamodes where all-out signal strength is not as important. Running 10W or so, you will not be the loudest station on the band, especially with SSB. But, if you are looking for a portable, lightweight, multi-band HF antenna, the Alpha Loop has a lot to offer. As long as you don't expect to work everything you can hear you won't be disappointed.

The design of the Alpha Loop has evolved since this review and in 2020 was rated at 100 watts. It costs US $399 (approx £309 at the February 2020 exchange rate) plus P&P and it can be ordered from the Alpha Antenna website at www.alphaantenna.com

THE SOTABEAMS BAND HOPPER III PORTABLE ANTENNA

Reviewed by Mike Richards, G4WNC
RadCom, October 2013

One of the ways to beat the RF noise pollution on the HF bands is to get away to a remote location and try some portable operation. The SOTABEAMS 'Band Hopper' antenna is designed to handle your portable antenna requirements in a neat, easy-to-erect, multi-band system. Of course, the system will also appeal to those involved in activating sites for the SOTA awards (Summits On The Air) or those who want to operate whilst on holiday.

The combination tested here is the Band-Hopper III antenna with the SOTAPole 7m mast. The SOTA Pole is a 7m-long, seven section, collapsible,

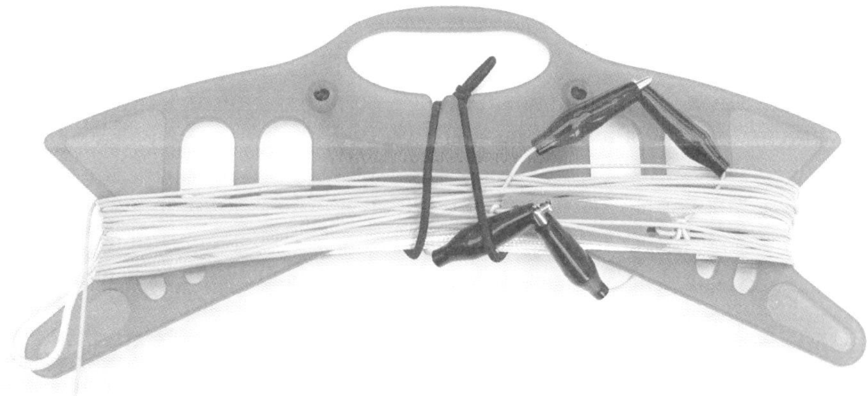

Winders used with the Band Hopper III.

fibreglass tube tapering from 35mm at the base to 2.1mm at the top. When collapsed into its carry case, the pole is just 1.14m long and weighs 650g.

The antenna is the Band Hopper III linked dipole antenna. As you can guess from the name, this is a multi-band system and there are a number of band combinations available. For the review I was testing the 40-30-20 model that covers, yes you've guessed it, the 7MHz, 10MHz and 14MHz amateur bands. The antenna is supplied fully assembled and ready to go with the guys and feeder pre-attached to the antenna elements. The feeder is 10m long and uses 50Ω RG-174 cable terminated in a crimped BNC plug.

The antenna elements and feeder are wrapped on to three string winders similar to those used for kites (see photo opposite). In addition to keeping the wires tidy, the winders can be used to secure the antenna in rocky terrain by wedging them behind a rock, or on heavy snow coverings they could be used as snow anchors.

The cable winders and the three ground pegs fit inside the compact soft carry case and the complete system weighs just 460g (excluding the pole). That makes a complete multiband HF antenna system that weighs just 1150g when you include the pole and case.

The design is extremely simple and the multi-band linked dipole is simply a half-wave dipole for 40m that is broken at the lengths for 30m and 20m half wave dipoles (see **Fig 6.2**). Each break in the antenna is bridged by a spacer and a pair of crocodile clips. These allow you to alter the length of the dipole by either joining or disconnecting the crocodile clips. The result is a half-wave dipole on each of the bands specified. When used with the SOTA pole, the antenna is erected as an inverted-V with the feed point at the top of the mast and the end points a few feet above ground.

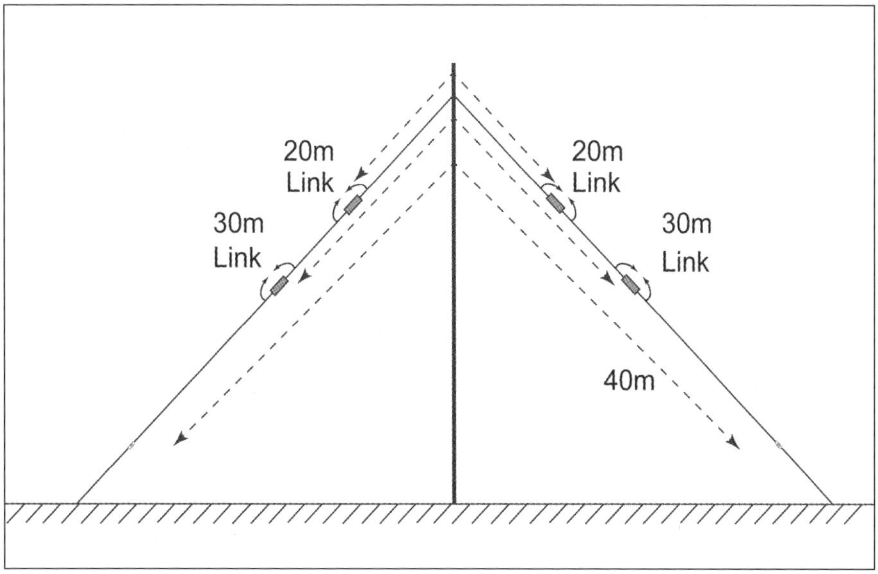

Fig 6.2: Band Hopper linked dipole configuration.

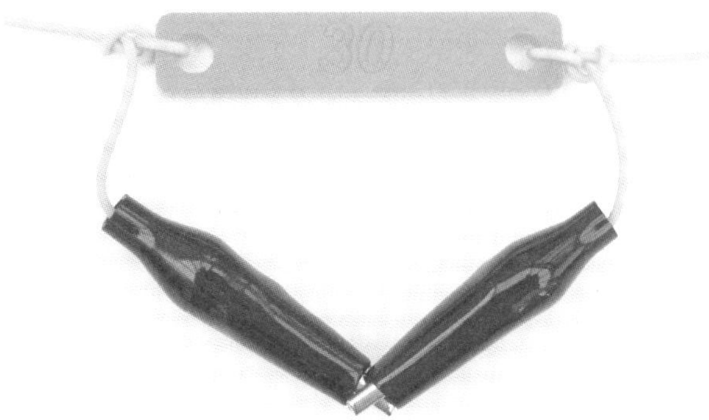

Close-up of the crocodile clip band selection links.

Although the pole only provides just over 6m of elevation to the feed point, the trick is to operate at a good location where you can take advantage of a naturally good take-off. We don't have any particularly high hills around us, so I decided to head for a site with a clear take-off to all points. I also wanted to find a site where I could park close to the antenna as I was intending to use the car's battery to power the station. The chosen site was just off the A31 on the old wartime Stoney Cross airfield. This is a large flat area at 356m above sea level with plenty of parking spaces close to the grassy areas.

Once parked up, the first job was to extend the pole and lock the sections together by twisting them. The pole was then placed on the ground with the top end pointing into the wind. The next stage was to insert the top section of the pole through the hole in central antenna spacer. The hole was just right for the top section to pass through but not the second section. The antenna wires were then set out on the ground at 90° to the pole using the winders to release the wire. I found that by holding the winder with one finger whilst walking, the wire unwound easily. The next step was to secure

The Band Hopper III mounted on the 7m SOTA pole in the New Forest.

the pre-attached guy lines to the pegs. To do this, I moved in about 2m from the end of each leg and inserted the peg and attached the guy string using the pre-tied loop. Once the two antenna ends had been secured, the instructions suggest moving the slack back to towards the centre of the antenna. At this point you need to decide which band you want to use and open or close the crocodile clips as necessary. The feeder was then unravelled from the third winder followed by the back-stay.

I was now ready to erect the antenna. With the back stay in one hand, I lifted the pole with the other from a point about 2m from the ground end and swung it vertically into place. I then moved the base around to get pole as near to vertical as possible and walked with the back stay to fix it to a third tent peg; job done. I think it took me longer to write this description than it did to erect the antenna! It really does only take about 5 to 10 minutes maximum to erect. The great point about the design is you can erect it single-handed even in quite windy conditions. Once erected, the feeder length was just right for operating portable beneath the antenna. I thought changing bands might be a bit tricky as you have to lower the antenna to reach the clips. However, my pessimism was misplaced and it was very easy to swap over. To make the change I released the back guy and gently lowered the antenna, adjusted the crocodile clips and gently raised it back to the operating position. I did this several times during my operating spell and it went without a hitch every time, taking only a couple of minutes to change.

For the review, my station comprised the new Zeus ZS-1 digital transceiver coupled to a laptop and feeding directly into the Band Hopper III feeder. Before connecting the Zeus, I checked the SWR of the Band Hopper that showed 1.3:1 for the range 7.000MHz to 7.190MHz and 1:1 for 7.060MHz to 7.120MHz. The SWR results were similar for the other bands and fine for direct connection of the Zeus. The first thing I noticed was the extremely low noise floor, which was around –120dBm (S1) on 7MHz in the afternoon. Back at the home QTH the noise floor was –85dBm (S7) at the same time of day – that's 35dB worse and a jolly good reason for operating portable!

The antenna has a maximum power rating of 125W but in the spirit of portable operation, I ran QRP with just 5 watts into the antenna. The set-up worked extremely well for me and I made contacts into Ireland and Germany with good signal reports. As the afternoon went on, the wind was increasing but the low drag of the antenna allowed it to remain stable.

I was very impressed with the Band Hopper antenna system. Whilst it is based on a simple half-wave dipole, the construction makes for a very compact, lightweight and easy-to-use antenna system that can be erected single-handed in just a few minutes. The instructions were also accurate and easy to follow. My thanks to SOTABEAMS for the loan of the review equipment.

The Band Hopper III is available direct from SOTAbeams (www.sotabeams.co.uk) and costs £69.95 inc VAT. It comes complete with the 10m of RG-174 feeder, a current balun, a guying system for a lightweight telescopic pole, three wire winders, three lightweight aluminium alloy pegs and a nylon carry bag. There are also two-band and four-band 'Band Hopper' antennas available at £59.95 inc VAT and £79.95 inc VAT respectively. The 7m heavy-duty SOTA Pole is £69.95.

Note: All the prices listed for commercial portable antennas were correct when this chapter was edited in January 2020 — Ed.

REFERENCES

[1] Common mode ferrite chokes 'cookbook' by K9YC: http://audiosystemsgroup.com/RFI-Ham.pdf

[2] Moonraker fibreglass mast: moonraker.eu (click on 'Antennas' then 'Masts' etc).

[3] VPA Systems products are sold by Teltad in Poland (www.teltad.pl).

[4] Guide to folding and unfolding the Folding Antenna by Lars, PH0NO: www.ph0no.net/p/ph0no-projects.html#project3

[5] *EZNEC:* www.eznec.com

Index to Antenna Articles